George Henry Napheys

Modern Therapeutics

A Compendium of Recent Formulae and Specific Therapeutical Directions. Second

Edition

George Henry Napheys

Modern Therapeutics
A Compendium of Recent Formulae and Specific Therapeutical Directions. Second Edition

ISBN/EAN: 9783337812522

Printed in Europe, USA, Canada, Australia, Japan

Cover: Foto ©berggeist007 / pixelio.de

More available books at **www.hansebooks.com**

:MODERN

THERAPEUTICS:

A COMPENDIUM OF

RECENT FORMULÆ

AND SPECIFIC

THERAPEUTICAL DIRECTIONS.

By GEO. H. NAPHEYS, A. M., M. D.

One of the Editors of the Half-Yearly Compendium of Medical Science; **Late**
Chief of Medical Clinic of Jefferson Medical College; Member of the
Philadelphia Co. Medical Society; Corresponding Member
of Gynæcological Society of Boston, etc., etc.

Second Edition--Revised and Improved.

PHILADELPHIA:
S. W. BUTLER, M. D., 115 SOUTH SEVENTH STREET.
1871.

PREFACE.

A year ago I began the publication in the MEDICAL AND SURGICAL REPORTER of a "*Therapeutical Bulletin.*" My design was to collate from recent medical periodicals, monographs, and systematic treatises, the utterances of experienced practitioners in relation to therapeutics. The favorable reception accorded to these articles induced me to prepare this book.

I am well aware that many sins of omission may be charged against me. It is simply impossible to treat of every disease, on the nosological list, within the limits of the present volume. These omissions I will endeavor—if my design meet with encouragement—to supply at some future time.

I have given the most space to those affections in which treatment has been found of most avail. In discussing such diseases as Cholera, Locomotor Ataxia, etc., in which therapeutical resources are, as yet, of little value, I have noted but a few of the best of the remedies which have been suggested

I think I can claim for this compilation the merit of novelty in object and arrangement. It contains not merely "recent formulæ," but "specific therapeutical directions," and, to some extent, the philosophy thereof, in the management of disease. All previous collections of therapeutical facts have been arranged with reference to the articles of the Materia Medica. The nosological plan here adopted, is I believe, the most convenient for the busy practitioner. It enables him to turn at once to the therapeusis of a disease.

Many of the formulæ I have obtained directly from the authors. To Dr. J. M. Da Costa, in particular, I wish to acknowledge my obligations for much valuable, and hitherto unpublished, matter.

Philadelphia, January, 1870.

NOTE TO SECOND EDITION.

This book has been out of print for some months. The first edition—a very large one—was exhausted in eight months. The absence of the author in Europe has delayed the appearance of the second edition, which is now offered to the profession in an enlarged and very much improved form.

Publisher.

January, 1871.

ERRATA.

R. 100, p. 39, should be under *Tetanus* instead of *Hysteria*.
Page 121, for Cardiac Region read Cardiac Neuroses.

TABLE OF CONTENTS.

MODERN
THERAPEUTICS.

I. DISEASES OF THE NERVOUS SYSTEM.

INSOMNIA.

The treatment of this form of nervous affection is given as
follows by

C. E. BROWN-SÉQUARD, M. D., F. R. S., ETC.

1. ℞. Potassii bromidi,   ℥ss.
     Aquæ cinnamomi,   f.℥ij.   M.

Sig.—Dessertspoonful a quarter of an hour before the last
meal, and the same dose, or three teaspoonfuls, repeated
at bed-time, for adults.

Excepting when pain is one of the causes preventing sleep,
(in which case the alkaloids of opium, aconite, or hyoscyamus
should be employed,) Dr. BROWN-SÉQUARD has found that
this remedy has a most wonderful power to produce a quiet
and refreshing sleep, without any drawbacks. In some cases
it is necessary to increase the dose of the bromide, and to
give also a small dose of narceine or codeine an hour before
bed-time. In those affections in which the bromide of potassium is not powerful enough as a sleep-inducing agent, a
warm bath of four, five, or six hours' duration is often successful.*

---

*Lectures on the Diagnosis and Treatment of Functional Nervous Affections.
By C. E. BROWN-SÉQUARD, M. D., F. R. S., Philadelphia, (Lippincott) 1868: p. 35.

1

SUPPOSITORY.

J. M. DA COSTA, M. D.

2. ℞. Assafœtidæ, gr. x.
Extracti hyoscyami, gr. iij—v. M.
Fiat suppository.
To be introduced at night, to quiet restlessness and induce sleep where it is not desirable to give opiates.

(GRAVES mentions that in persons of irritable and nervous disposition he has found musk or assafœtida given more or less frequently during the day effectual in procuring sleep at night.*)

3. ℞. Pulveris digitalis, gr. iij.
Extracti hyoscyami,
Camphoræ, aa gr. xij. M.
For twelve pills.
One to be taken at night.

Dr. DA COSTA has found in reference to the soporific and anodyne properties of narceine, that it appeared, in doses in which morphia is prescribed, totally destitute of either; and in larger doses uncertain, and often palpably inert. It does not allay irritation. (*vide* Pennsylvania Hospital Reports for 1868.)

PROF. WILLIAM A. HAMMOND, M. D., ETC.

The principles which should prevail in the treatment of wakefulness may be arranged into two classes.

1st. Those which by their tendency to soothe the nervous system, or to distract the attention, diminish the action of the heart and blood vessels, or correct irregularities in their function, and thus lessen the amount of blood in the brain.

2d. Those which directly, either mechanically or through a specific effect upon the circulatory organs, produce a similar effect.

*Functional Nervous Disorders; C. Handfield Jones, p. 176.

In slight cases, the measures belonging to the first class often prove effectual. Among them are music, monotonous sounds, gentle frictions of the surface of the body, soft undulatory movements, the repetition by the insomnolent of a series of words till the attention is diverted from the existing emotion which engages it, and many others of similar character. In persistent insomnia, however, these are nugatory.

Chief among the means embraced under the second head are those which tend to improve the general health of the patient.

In regard to food, while it is an error to suppose, as is generally done, that a moderately full meal, eaten shortly before bed-time, is necessarily productive of wakefulness, there is no doubt that this condition is induced by an excessive quantity of irritating or indigestible food. A hearty supper of plainly cooked and nutritious food rather predisposes to sleep. This is due to the fact that the process of digestion requires an increased amount of blood in the organs which perform it, and, consequently, the brain receives a less quantity. This hypnotic effect is neutralized, however, when the food is immoderate in amount or irritative in quality, as it then either by the pressure upon the abdominal vessels, or through a reflex action on the heart, augments instead of diminishing the quantity of blood circulating in the brain. Attention should therefore be paid to the diet of the insomnolent. As a rule people are under-fed. This is especially true of women. The tone of the system is thus lowered and local congestions of different parts of the body are produced. If the brain be one of these, wakefulness results.

Most of the cases of insomnia in women are of the passive variety, and require not only nutritious food, but *stimulants*. Whisky is generally to be preferred to brandy and many kinds of wine. Nothing can be better as a good stimulant, and at the same time tonic, than *Tarragona wine*, drank at dinner to the extent of a glass or two. Next must be ranked good *lager beer*.

There are cases in which *coffee* induces sleep. Our author mentions several in which passive wakefulness was entirely and speedily cured by a cup of strong coffee taken for three or four nights in succession at bed-time. In females of languid circulation and a consequent tendency to internal congestions, it is particularly useful.

The employment of stimulants is only of service in the asthenic or passive form of insomnia; in the sthenic or active form they would of course increase the difficulty.

Physical exercise in the open air, extended to the point of inducing a slight feeling of fatigue, is productive of good effects.

The *warm bath* calms nervous irritability and determines blood from the head. Putting the feet in water of the temperature of 100° F. will often induce sleep, particularly in children, after other means have failed.

Cold water (32° F.) applied directly to the scalp has a good influence in those cases in which the individual is strong, the heart beating with force and frequency, and the mental excitement great. It is not admissible in the asthenic form of wakefulness.

Among the purely medicinal agents, *bromide of potassium* holds the first rank. It diminishes the amount of blood in the brain, and allays any excitement which may be present in the sthenic form of insomnia. The flushed face, the throbbing of the carotids and temporals, the effusion of the eyes, the feeling of fullness in the head, all disappear as if by magic under its use. It may be given in doses of from ten to thirty grains—the latter quantity is seldom required, but may be taken with perfect safety in severe cases. (See F. 1.)

4. R. Zinci oxidi,   ℈ij.
        Confectionis rosæ,   q. s.
    For twenty pills. One ter die, the last dose being taken just at bed-time.

Our author seldom employs *opium;* there are cases of insomnia, however, in which its influence is decidedly beneficial.

5. R. Tincturæ hyoscyami,   f.ʒij.
    From one to two teaspoonfuls in water at bed-time.

Especially indicated in those cases which are accompanied by great nervous irritability. A good preparation of this drug is difficult to obtain. It possesses no advantages over bromide of potassium, to which it is not equal in any respect.

Our author has nothing to say in commendation of valerian, assafœtida, and other anti-spasmodics. Tonics, especially iron and quinine, are almost always useful, even in the active forms of the affection.

In insomnia dependent upon severe and long-continued mental exertion, all means will fail if the individual will not consent to use his brain in a rational manner. Proper intervals of relaxation must be insisted upon, and in some cases mental rest. Travel is always of the greatest advantage in such cases.*

HYPODERMIC INJECTION.

ANTOINE RUPPANER, M. D., FELLOW MASS. MED. SOC.

6. R. Tr. hyoscyami.   gtt. x—xx.
        For a single injection.
7. R. Tr. cannabis indicæ,   gtt. x—xx.
        For one injection.†

---

*Sleep and its Derangements. By William A. Hammond, M. D. J. B. Lippincott & Co., Phila., 1869, pp. 278, et. seq.
†Hypodermic Injections in the Treatment of Neuralgia, Rheumatism, Gout, and other Diseases; p. 31.

THOMAS HAWKES TANNER, M. D., F. L. S., ETC., LONDON.*

8.　℞.　Extracti stramonii,　　　　　gr. iij.
　　　　 "　hyoscyami,　　　　　　　gr. xviij.
　　　　 "　lupuli,　　　　　　　　　℈ij.　　　　M.
　　For twelve pills.
　　One to be taken every four hours until relief is obtained, in
　　chronic disorders attended with suffering ; in diseases of
　　the nervous system accompanied with pain and restless-
　　ness; and in the dyspnœa of phthisis and emphysema.

Our author has seen good results in cases of wakefulness,
particularly when there is any debility, from a tumblerful of
port wine negus, or of mulled claret, or of white wine whey,
taken the last thing at night. Where the skin is hot and dry
a glass of cold water appears to be useful.

When there is any physical cause it must be removed. If
the bowels are constipated or the excretions unhealthy, laxatives and alteratives will be required. Patients afflicted with
heart-burn should take three or four bismuth lozenges before
retiring to rest. If sedative drugs are necessary resort should
first be had to henbane, hops, Indian hemp, or conium. When
stronger drugs are needed the following may be prescribed :

9.　℞.　Morphiæ muriatis,　　　　　gr. i–ij.
　　　 Spiritus chloroformi,
　　　 Tr. cardamomi compositæ,　　aa f.ʒj.　　M.
　　Dessertspoonful at bed-time ; or
10.　℞.　Extracti opii,　　　　　　　gr. iij.
　　　 "　hyoscyami,　　　　　　　gr. xxiv.　　M.
　　For six pills.　One at bed-time.

Frequently the exhibition of opiate enemata or suppositories
is preferable to the use of this drug by the mouth. The following may be employed:

11.　℞.　Tincturæ opii,　　　　　　♏ xx-xxx.
　　　 Mucilaginis amyli.　　　　　f. ʒij.　　M.
　　For one enema.
12.　℞.　Pulveris opii,　　　　　　　gr. i–ij.
　　　 Saponis duri,　　　　　　　gr. x.　　M.
　　For a suppository.

---

*Practice of Medicine, Am. Ed., (L. & B.) 1866 ; pp. 307, 76).

EDWARD JOHN TILT, M. D., M. R. C. P., ETC., LONDON.*

13. ℞. Extracti hyoscyami, gr. xxiv.
 " cannabis indicæ, gr. iij. M.
For twelve pills.

One or two to be taken at night, or oftener. But Dr. TILT gives Indian hemp in one-grain doses, as soon as he finds it agrees, and sometimes in larger doses. If he desires a tonic as well as sedative effect, he orders

14. ℞. Extracti hyoscyami,
 Quiniæ sulphatis, aa gr. xij. M.
For twelve pills.

One to be taken every night. This is a preparation that he has often found to be well borne by women who could not bear large doses of any tonic ; some have continued to take it for months, not leaving it off during the menstrual period ; and it will not interfere with the action of any purgative that may be required.

CHARLES WEST, M. D., F. R. C. P., LOND., ETC.†

DR. CHAS. WEST states that the value of tincture of hyoscyamus as a sedative in the diseases of children can scarcely be too highly estimated. He orders,

15. ℞. Tr. hyoscyami, ℳxviij.
 Syrupi, f.ℨiij.
 Aquæ, f.ℨix. M.

Dessertspoonful every six hours for a child a year old.

To this mixture there may be added, if there is much peevishness.

16. ℞. Potassæ bicarbonatis,
 Acidi citrici, aa gr. xx.

Also, if the stomach be not irritable,

 Vini ipecacuanhæ, ℳ xij.

---

*Hand-Book of Uterine Therapeutics, Am. Ed., 1869, p. 333.
†Lectures on the Diseases of Infancy and Childhood. Am. Ed.,1856. p. 45.

DR. FORBES WINSLOW.

Our author gives a hint which is worth remembering with regard to the employment of sedatives generally. This is, that cases which are intractable to separate remedies will yield to a judicious combination of several. This is probably the secret of the success of the nostrum, chlorodyne.

A placebo, as a bread pill, is often a powerful sedative, if the patient can be inspired with faith in its efficacy. Dr. LAY-COCK mentions a case in which sleep, after taking such a pill, was so long as to excite alarm.

(The treatment of Insomnia caused by pain is considered under the heads of Neuralgia, and of Rheumatism.)

HEADACHE.

J. M. DACOSTA, M. D.

In *congestive headache* the use of saline cathartics is of service. Also, a mustard foot-bath every night. The application of a hot salt bag to the back of the neck often affords relief. This form of headache is frequently associated with cardiac enlargement.

PROF. AUSTIN FLINT, M. D., ETC., NEW YORK.

PERIODICAL HEADACHES,

As regards successful treatment, belong among the opprobria of medical art. If patients be not unpleasantly affected by opiates, an attack may sometimes be warded off or its severity lessened by a full dose of this drug or one of its alkaloids. The carbonate of ammonia and a saline purgative are sometimes effective at the commencement of an attack. Various

palliative measures may be resorted to, such as inhalation of
chloroform, evaporating lotions to the head (alcohol, spirits,
vinegar, ether), etc. In some cases a towel or napkin, wrung
out in water as hot as can be borne, and wound around the
head, is more efficient than cold applications. Warm stimulating pediluvia, strong coffee or tea, and the application of
the galvanic or the electro-galvanic current are useful in
some cases. During the intervals the remedies which are
sometimes of service by way of prophylaxis are, nux vomica
or strychnia in small doses, arsenic, small doses of quinia,
belladonna, and the preparations of zinc, more especially the
valerianate. They may be tried in succession. Hygienic
measures are important, and the avoidance of everything,
which experience shows in individual cases, to act as exciting causes.

NERVOUS HEADACHES.

PROF. WILLIAM A. HAMMOND, M. D., ETC., NEW YORK.

17. ℞. Zinci oxidi,   ℈ij—℈v.
      Confectionis rosæ,   q. s.   M.
      Divide into xx pills.
      One to be taken three times a day after meals.

This formula is of great value. The minimum dose (gr. ij.)
should be commenced with, gradually increasing to the
maximum (gr. v.) if necessary.

18. ℞. Extracti nucis vomicæ,   gr. v.
      Ferri redacti,   ℈j.
      Quiniæ sulphatis,   gr. x.
      Syrupi,   q. s.   M.
      Divide into xx pills.
      One to be taken three times a day after meals.

Nux vomica is preferable to strychnia.

19. ℞. Bismuthi subcarbonatis,   ℈ij.
      Confectionis rosæ,   q. s.   M.
      Divide into xx pills. One after each meal.

These pills will often take the place of those of oxide of zinc (F. 17). They are particularly useful when there is gastric disturbance.

Bromide of Potassium is serviceable when the nervous system has been irritated; when exhausted it does harm.

Bromide of Ammonium is similar to the bromide of potassium in its action, but the dose need not be so large. Our author often uses both combined.

Opium and its preparations are rarely of value in this disorder. If used, the hypodermic method is the best.

Narcein has, Dr. H. still thinks, a decided hypnotic effect when given in large doses.

Phosphorus is beneficial in all the forms of nervous headache. It is, however, difficult of administration and leaves an unpleasant odor about the person. The best results are obtained from the following method of administration:

  20. R. Acidi phosphorici diluti, f.℥vj.
     Syrupi phosphatis compositi, f.℥iij. M.
  A dessertspoonful, in water, three times a day.

Arsenic as a nerve tonic stands next in value to zinc. Granules of arsenious acid (gr. 1-40) are preferable to FOWLER'S solution. *Galvanism* is highly praised by some and severely condemned by others in this affection. The brain cannot be acted upon to any considerable extent by the induced current or by reflex action. Our author advises always the *constant current;* being careful to avoid too great intensity lest amaurosis be produced.*

   DR. GEORGE KENNION, OF HARROWGATE, ENG.

  21. R. Carbonis bisulphidi,   f.℥ij.
  As a local application in neuralgic, periodical and hysterical headache, and even in many cases of dyspeptic cephalalgia.

Half Yearly Compendium of Medical Science, July 1868. p. 67.

About two drachms of the bisulphide of carbon is poured upon cotton wool, with which a small glass stoppered bottle is half filled. The mouth of the bottle is applied *closely* to the temple or behind the ear, or as near as possible to the seat of pain, and so held from three to five or six minutes. In a minute or two a sensation as of several leeches biting the part is felt, and in three or four minutes more the smarting and pain become rather severe, but subside almost immediately after the removal of the bottle. It is very seldom any redness of the skin is produced. The effect of the application is generally immediate; it may be repeated, if necessary, three or four times a day. The sedative vapor of the bisulphide is probably absorbed through the skin, and acts upon the superficial nerves of the part to which it is applied.*

THOMAS HAWKES TANNER, M. D., F. L. S., ETC., LOND.

22. ℞. Quiniæ sulphatis, gr. xxiv.
   Pulveris rhei, gr. xxxvj.
   Glycerinæ, q. s. M.
   Divide into twelve pills, and order one to be taken at night.

Often of service in curing bilious headaches; the patients also taking daily exercise in the open air, and avoiding too much sleep.

23. ℞. Acidi nitro-muriatici diluti, f.ʒij.
   Strychniæ, gr. ¼—½
   Spiritûs chloroformi, f.ʒvj
   Tincturæ zingiberis, f.ʒiij.
   Aquæ, q. s. ad. f.ʒiij. M.
   A teaspoonful in water three times a day, *in nervous head-ache.*

Holding the arms high above the head produces a marked effect upon the cerebral circulation, and will frequently relieve

* *Medical Times and Gazette*, July 1863, p. 77.

the severity of that peculiar morning headache, with which some persons constantly awake.

Compression of the temporal arteries with a couple of pads and a bandage may sometimes be of service.

Cold lotions, eau de cologne, etc., to the head, dry cupping, or blisters, or setons to the nape of the neck; the removal of decayed teeth or stumps from the mouth, and change of air, are occasionally indicated.

24. ℞. Zinci valerianatis,   gr. xij—xxiv.
Extracti belladonnæ,   gr. iij.—vj.
Extracti gentianæ,   gr. xxiv.   M.
Divide into twelve pills. One to be taken three times a day.

Useful in *hysterical headache* especially when there is habitual constipation.

25. ℞. Zinci phosphatis,   ℨj—ij.
Acidi phosphorici diluti,   f.ℨjss.
Tincturæ cinchoniæ,   f.ℨvj.
Aquæ menth. pip.,   q. s. ad f.ℨiij.   M.
Tablespoonful in a half wine glass of water three times a day, *in hysterical headache associated with debility.*

HENRY G. WRIGHT, M. D., M. R. C. P., ETC.

26. ℞. Tincturæ capsici,   f.ℨij.
Liquoris ammoniæ acetatis,
Tincturæ aurantii corticis,
Syrupi aurantii corticis, aa f.ℨvj.
Aquæ,   f.ℨss.   M.
Dose—a tablespoonful. To relieve the headache that ensues after *inebriety,* etc.

27. ℞. Linimenti chloroformi,
Linimenti belladonnæ,   aa f.ℨiss.
Tincturæ opii,   f.ℨj.   M.
For external application, *in rheumatic headaches.*

Mustard plasters applied to the neck are also exceedingly

useful as a means of counter-irritation. In such cases (rheumatic headache) the following aperient is of advantage, viz:

28. ℞. Pilulæ colocynthidis comp., gr. xv.
Extracti colchici acetici, gr.iij.
Olei carui, ℳj. M.

Divide into four pills. Two to be taken at bed-time, and one on consecutive nights. These pills should be followed, in persons of a costive habit, by a morning purgative, as follows :

29. ℞. Magnesiæ, ℈iv.
Liquoris potassæ, ℳxlv.
Extracti sennæ fluidi, f.ʒij.
Syrupi zingiberis,
Tincturæ aurantii corticis, aa f.ʒss.
Aquæ f.ʒj. M

Dose—a tablespoonful.

The administration of an alkaline medicine containing potash, if continued with regularity, will generally be followed by rapid amelioration of the pain and tenderness. The following may be used :

30. ℞. Potassæ carbonatis, ℈iv.
Potassæ chloratis, ʒiss.
Tincturæ cinnamomi,
Tinct. aurantii corticis, aa f.ʒvj.
Syrupi aurantii corticis, f.ʒjss. M.

A dessert spoonful to be taken twice or three times a day·

If imprudent exposure to cold has produced an aggravation of the headache, and particularly if the patient be subject to catarrh it is advisable to administer a sudorific at bed time such as

31. ℞. Pulveris ipecacuanhæ comp., gr. xij.
Pulveris camphoræ,
Pulveris guaiaci, aa gr. iv. M.

For one powder, to be taken about bed-time.

In *gouty* headaches colchicum may be employed with greater freedom than in ordinary gout, care being taken that the bowels are freely open during its administration. An actual attack is best relieved by a brisk aperient, (F. 28) followed by an effervescing mixture, containing an excess of potash, viz.:

32. ℞. Potassæ carbonatis,   ℈iv.
          Ammoniæ carbonatis,   ℈ij.
          Tincturæ serpentariæ,   f.ℨss.
          Aquæ camphoræ,   f.ℨiijss.   M.

Two tablespoonsful to be added to a tablespoonful each of water and lemon juice, and to be taken effervescing twice or three times a day.

In the treatment of *plethoric headaches* the employment of medicines should as far as possible be dispensed with. They should only be resorted to when the necessities of business prevent, or the solicitations of indolence interfere with a strict control over the diet and regimen. In these cases a saline diuretic should be ordered, such as

33. ℞. Potassæ acetatis,   ʒij.
          Potassæ nitratis,   ʒj.
          Spiritûs juniperi compositi,   f.ʒxj.
          Aquæ menth. pip.,   q. s. ad. f.ʒiv.   M.

A teaspoonful twice a day together with an occasional aperient at night, viz:

34. ℞. Pil. colocynthidis comp.,   gr. l.
          Saponis castilliensis,   gr. ix.
          Olei anethi,   ℨ ij.   M.

Divide into twelve pills; two to be taken at bed-time and followed by a Seidlitz powder in the morning.

Persons subject to plethoric headaches should not partake of animal food more than once a day; should never indulge the appetite to satiety; should avoid beer, spirits, coffee and all stimulating beverages; should bathe the head freely at night and lie with it elevated on a hard pillow during sleep; should have an airy bed-room and rise so soon as fairly awake, for otherwise activity of the thoughts in a recumbent position will congest the head and cause it to ache.

When the patient has been exposed to cold and the headache comes on at night, with the head hot and the skin harsh and dry the following sudorific, taken at bed time, is often of great service, the body being kept warm during its action.

35. R. Pulveris antimonii et potassæ
  tartratis,   gr. 1–6.
  Pulveris Jacobi veri,   gr. v.
  Pulveris potassæ nitratis,   gr. x.   M.
  To be taken at night.

In the plethoric headache of pregnancy, relief is afforded by the use of saline medicines, as

36. R. Magnesiæ sulphatis,
  Sodæ sulphatis,   aa ʒj.
  Acidi sulphurici diluti,   f.ʒij.
  Tincturæ cardamomi comp., f.ʒjss.
  Syrupi aurantii corticis,   f.ʒss.
  Aquæ cinnamomi,   f.ʒj.   M.
  A dessertspoonful twice a day.

Fluids should be avoided as far as possible. Sea air and sponging the body with tepid salt water generally prove beneficial.

In the treatment of *congestive headaches*, the aperients so frequently required should be cordial and saline, such as

37. R. Extracti sennæ fluidi,   f.ʒijss.
  Magnesiæ sulphatis,   ʒij.
  Acidi sulphurici aromatici, f.ʒij.
  Syrupi aurantii,   f.ʒj.
  Infusi rhei,   f.ʒijss.   M.
  Dose—a tablespoonful in the morning.

There are no medicines so invariably useful in cases of congestive headache, attended with debility, as the preparations of iron. If the patient be of stout phlegmatic habit, the tonic may be combined with a cordial and saline, according to the following formula :

38. R. Ferri sulphatis,   gr. xxxij.
  Magnesiæ sulphatis,   ʒx.
  Acidi sulphurici diluti,   f.ʒij.
  Tinct. cardamomi comp.,   f.ʒij.
  Syrupi,
  Aquæ pimentæ,   aa f.ʒj.   M.
  A dessertspoonful in water twice a day.

39. ℞. Tincturæ ferri chloridi,   f.ʒij.
         Acidi muriatici diluti,   f.ʒss.
         Tincturæ cinnamomi,   f.ʒjss.
         Syrupi,
         Aquæ cinnamomi,   aa f.ʒvj.   M.

A dessert-spoonful in water twice a day, about an hour
after food.

In the congestive headache of females, past the middle period
of life, especially when these headaches accompany alterations
of the whole system, at the great climacteric period.

In the treatment of *dyspeptic headaches*, when the pain
comes on directly after a meal, and when it can be traced to
indigestible articles of food, and the patient is tolerably strong,
an emetic is useful.

40. ℞. Pulveris ipecacuanhæ,   gr.xxv.
         Ammoniæ carbonatis,   gr. v.
         Aquæ menthæ viridis,   f.ʒiss.   M.

Take at one dose and follow by some warm fluid.

Where the pain ensues some hours after taking food, a warm
draught, with the following formula, is generally beneficial :*

41. R. Pulveris rhei,   Ðij-s.
         Magnesiæ carbonatis,   Ðij.
         Spiritûs ammoniæ aromatici,   f.ʒij.
         Syrupi zingiberis,   f.ʒss.
         Aquæ menthæ piperitæ, q. s. ad. f.ʒij.   M.

A tablespoonful in water.

Such a headache may often be warded off by the following :

42. ℞. Pulveris rhei,   gr. xviij.
         Pulveris capsici,   gr. v.
         Sodæ carbonatis exsiccatæ,
         Pulveris aloes,
         Saponis castilliensis,   aa. gr. xij.   M.

Divide into xij pills.

One to be taken before the meal as a dinner pill.

___

*Hea laches: Their causes and their cure. Am. Ed., p. 88.

43. ℞. Pilulæ hydrargyri,
   Pilulæ rhei compositæ,   aa gr. iv.
   Extracti hyoscyami,   gr. ij.   M.

Divide into two pills.

To be taken at night, in cases of headache depending upon *dyspepsia*, or

44. ℞. Pulveris ipecacuanhæ,   gr. j.
   Pilulæ colocynthidis comp.,   gr. vij.
   Extracti gentianæ,   gr. ij.
   Olei carui,   ℔ ½.   M.

Divide into two pills. To be taken at bed time.

When the system is debilitated it is often more advisable to employ the following, in place of the night pills:

45. ℞. Infusi sennæ,
   Infusi rhei,   aa f.ℨv.
   Tincturæ cardamomi comp.,   f.ℨj.
   Syrupi,   f.ℨss.   M.

To be given as a draught in the morning.

Great benefit is generally derived from the combination of a bitter with an alkali, as in

46. ℞. Sodæ carbonatis,   ℨij.
   Spiritûs ammoniæ aromatici,   f.ℨvj.
   Tincturæ aurantii,
   Syrupi aurantii,   aa f.ℨiij.
   Tincturæ gentianæ comp.,   f.ℨjss.

A dessertspoonful twice a day.

If the stomach be very irritable, with excessive flatulence, the following mixture is more suitable:

47. ℞. Bismuthi subnitratis,
   Sacchari albi,
   Pulveris acaciæ,   aa ℨj.
   Tincturæ cardamomi compositæ,
   Tincturæ zingiberis,   aa f.ℨss
   Aquæ anethi,   f.ℨiijss.   M.

A tablespoonful taken twice a day.

2

The same medicine may be combined with soda and capsicum, thus :

> 48. R. Bismuthi subnitratis,
>   Sodæ carbonatis exsiccatæ, aa ℈ijss.
>   Pulveris capsici,   gr. viij.   M.
>   For eight powders. One to be taken twice a day, in
>   dyspepsia with much acidity, with loss of appetite
>   and general want of tone.

> 49. R. Argenti oxidi,   gr. xij.
>   Pulveris capsici,   gr. iv.
>   Extracti gentianæ,   gr. xxiv.   M.
>   Divide into eight pills. One to be taken twice a day.

In atonic dyspepsia, when the tongue is pale at the tips
and edges, and the system weakly, few recipes prove so invariably successful as the above.

> 50. R. Morphiæ muriatis,   gr. j.
>   Pulveris camphoræ,   gr. xxiv.
>   Mucilaginis gummi acaciæ,   q. s.   M.
>   Divide into six pills.
>   *In Intermittent Headache.*—One pill to be taken when the
>   headache is very intense. Five hours are to be allow-
>   ed to elapse before repeating the dose.

> 51. R. Quiniæ sulphatis,   ʒss.
>   Ferri carbonatis saccharatæ,   ʒj.
>   Pulveris aromatici,   ʒijss.   M.
>   Divide into xx powders.

> One to be taken twice or three times a day, commencing
>   with half a powder. Useful in *periodical headache* oc-
>   curring in delicate and weakly persons.

In the treatment of *Nervous Headache* and of cases of meg-
rims, the combinations of hyoscyamus with camphor, with
chloroform, or with a diffusible stimulant, generally affords
great relief. The following formulæ may be used :

> 52. R Extracti hyoscyami,
>   Pulveris camphoræ,   aa ℈ijss.   M.
>   Divide into xx pills.
>   Two to be taken when the pain is severe.

53. R. Chloroformi,   f.℥iss.
Tincturæ hyoscyami,
Tinct. cardamomi comp.,   aa f.ʒss.
Olei limonii,   ℥xvj.
Sacchari albi,
Pulveris acaciæ,   aa ʒss.
Aquæ camphoræ,   f.℥iij.   M.
Dose—a tablespoonful.

54. R. Tincturæ hyoscyami,
Spiritûs ammoniæ aromat., aa f ℥ss.
Syrupi aurantii,   f.℥j.
Aquæ menthæ piperitæ, f.℥ij.   M.
Dose—a tablespoonful.

55. R. Pulveris camphoræ,
Quiniæ sulphatis,   aa gr. x.
Extracti aloes,   gr. xij.
Extracti hyoscyami,   ʒss.
Mucilaginis gummi acaciæ, q. s.   M.
Divide into xviij pills. Two pills to be taken twice a day day, in cases of nervous headache when there is great debility and sluggishness of the system.

(For additional recipes, see Hemicrania.)

VERTIGO.

C. HANDFIELD JONES, F. R. C. P., LONDON, PHYSICIAN TO ST. MARY'S HOSPITAL, ETC.

56. R. Hydrargyri chloridi corrosivi, gr. j.
Glycerinæ,   f.℥j.
Tincturæ cinchoniæ compositæ, f.℥ij.
Olei menthæ piperitæ,   ℥xxv.   M.
A teaspoonful in a wineglassful of water three times a day.

In the *vertigo of old persons* which occurs, sometimes paroxysmally, as a single symptom, unassociated with any special state that can account for it, Dr. THOMAS HAWKES TANNER also recommends this treatment, by small doses of corrosive sublimate and bark, for the attacks of temporary dizziness to which the aged are liable.

ACUTE MENINGITIS.

WILLIAM AITKEN, M. D., EDIN., ETC.

When arachnitis arises from mechanical injuries, the treatment is generally by bleeding, active purgatives, especially by calomel and scammony, and by cold applications to the head.

In advanced life general blood-letting is rarely indicated, but in vigorous constitutions it is sometimes necessary. As a rule, local blood-letting is more safe and more beneficial, especially when aided by keeping the head well raised and by the constant application of cold water to the scalp or the occasional use of bladders filled with crushed ice. The bowels should be opened as rapidly as possible unless the patient is feeble, emaciated, or greatly exhausted. For this purpose employ

> 57. R. Hydrargyri chloridi mitis, gr. ij—iij.
> Extracti colocynthidis comp., gr. iv—v. M.
>
> Divide into two pills. Both to be taken at a dose, in cases accompanied with gout or Bright's disease, and followed in a few hours by a dose of salts and senna.

A fair proportion of nutriment must be given in the form of milk, strong beef-tea, sago, tapioca, or arrowroot; and the patient should bo kept in a quiet and darkened room. The more active symptoms being subdued, but not till then, a blister should be placed on the nape of the neck, if coma should ensue. If nervous irritability continues during convalescence, henbane or muriate of ammonia may be given. The bowels are to be kept open, and the strength supported by unstimulating nutriment. Tranquility of mind and body must be preserved.*

---

*Science and Practice of Medicine. 2d Am. Ed., vol. ii., p. 293.

DR. FELIX VON NIEMEYER, PROF. UNIVERSITY OF TUBIN-
GEN, ETC.

In the treatment of acute meningitis it is not generally
proper to bleed from the arm, but leeches may be applied to
the brow and behind the ears, and, if the strength of the patient
permit, the application may be repeated. The shaven head
may be covered with cold compresses, and an active purge of
calomel and jalap administered, viz. :

58. R. Hydrargyri chloridi mitis,   gr. ij.
       Extracti jalapæ,   gr. viij.   M.
    Divide into two pills and order both to be taken.

In the latter stages of the disease, if there be coma and
other signs of cerebral palsy, apply a large blister to the nape
of the neck and rub the following pustulating ointment on the
head :

59. R. Olei tiglii,   ℥xv.
       Adipis,   ℥ss.   M.
    One-fourth part to be rubbed into the skin every eight hours,
    until an abundant eruption is produced.

Still more efficacious than these derivatives are douche
baths, pouring cold water over the head from a pitcher held
some distance above it. The patient almost always recovers
consciousness as this is being done, but it must be repeated
at intervals of a few hours to secure a permanent result ;
with each successive employment the number of pitcherfuls
is to be increased. Frictions with mercurial ointment and
continued doses of calomel are much employed.

DELIRIUM TREMENS.

WILLIAM AITKEN, M. D., EDIN.

The two indications for treatment are: 1. The elimination of the poison ; 2. The sustenance of the patient during this period. Our author opposes bleeding and the administration of opiates or stimulants in large doses. The strength is to be supported by nutritious diets, such as yolk of eggs, soups, beef-tea, and egg-flip, in small quantities and often.

The danger in the first instance is from exhaustion, which is to be met by careful nursing. Opium may only be administered in protracted cases, and then never in doses larger than would be considered safe for a healthy person of the age and sex of the patient.*

J. WARING-CURRAN, L. K., AND Q. C. P. I., ETC.

60.   R.   Zinci oxidi,   Ʒj–ij.
           Confectionis rosæ,   q. s.   M.
For x pills ; one ter die.

When morphia or cannabis has done its duty in this disease, the after treatment by oxide of zinc is something to be observed rather than described ; the constant dread, restlessness, and disturbed sleep are quickly overcome by the bracing agency of the drug. In administering the oxide of zinc care must be taken not to give it upon an empty stomach, as it produces nausea and a dislike for the medicine.†

DR. G. M. JONES, OF JERSEY, ENGLAND.

61.   R.   Tincturæ digitalis,   f.℥jss.

A tablespoonful (f.℥ss) to be given at a dose, mixed with a little water. If the first be not sufficient, which, however, it generally proves to be, a second equally large is to be administered in about four hours. If a third dose be, in rare instances, required, it should not exceed a dessertspoonful (f.℥ij.)

---

*Science and Practice of Medicine. 2d Am. Ed. p. 779.
†London Lancet, Oct. 24th, 1868, p. 538.

Under the influence of this medication, it is stated, the pulse becomes fuller, stronger, and more regular, the skin grows warm and the cold clammy perspiration ceases. These effects are followed by a sleep of several hours' duration. No action on the kidneys nor any alarming symptoms are observed.*

Dr. T. HAWKES TANNER confirms the above statements.

This treatment answers best when the symptoms have assumed a resemblance to those of acute mania, and when there has not been much exhaustion.†

DR. LYONS, HARDWICKE HOSPITAL, DUBLIN.

62. R. Pulveris capsici. gr. xx–xxx
    Mellis rosæ, q. s. M
    Fiat bolus.
For one dose.

This usually suffices to produce quietude and sleep. In exceptional instances, however, a second and even a third dose is required before full tranquility is secured. The drug is well borne, and quiets the stomach in cases in which irritability and vomiting are present. Our author sums up his experience as follows:

1st. Capsicum is a valuable and reliable drug when opium fails or is for any cause contra-indicated. 2d. It is a safe drug for general employment in delirium tremens, and as such may be confidently recommended for general employment. 3d. It is not open to the objection which attaches to the continued use of opium which, when it fails to tranquilize and produce sleep, adds to the state of excitement, and when pursued beyond a certain limit may induce opium coma. 4th.

---

*Stille's Therapeutics. 3d Ed., vol. ii, p. 253.
†Practice of Medicine. Am. Ed., 1866: p. 247.

Capsicum has been employed in delirium of fever when opium has failed to cause sleep, and with marked success in certain cases.

As a member of the family of solanaceous plants, capsicum might a priori, have been expected to contain a narcotic principle. As yet the alkaloid in which it resides has not been isolated; but in some researches, conducted at the request of Dr. LYONS, M. ALPHONSE GAGES, a distinguished member of the chemical staff of the College of Sciences for Ireland, has found sufficient indications of its presence to warrant him in predicting its ultimate detection and isolation.. It will, Dr. Lyons expects, form a valuable boon to practical medicine when eliminated from the acrid oils of the capsicum fruit.*

SUNSTROKE.

DR. WHITEHILL, OF ST. LOUIS.

Our author has had a large experience with sunstroke, having seen as many as fifty cases in a single day during a forced military march in 1863.†

The treatment found most successful was cold to the head and chest, friction of the extremities and the internal administration of stimulants, such as brandy and ammonia. In his own case the nausea and vomiting were relieved by full draughts of strong green tea and Rhenish wine. In all cases a most important part of the treatment was to place the patient in the recumbent position in the shade, where there was a free circulation of air and at the same time disencumber him of everything that could in any wise interfere with either circulation or respiration. Under this treatment every case had recovered.

* *British Medical Journal*, Nov. 7th, 1869, p. 497.
† St. Louis *Medical Archives*, Sept. 1868, p. 401.

EPILEPSY.

C. E. BROWN-SÉQUARD, M. D., F. R. S., ETC.

63. R. Potassii iodidi,
    Potassii bromidi,   aa ℥j.
    Ammonii bromidi,   ℥ss.
    Potassæ bicarbonatis,   Ɉij.
    Infusi calumbæ,   f.℥vj.   M.

A teaspoonful before each of the three meals and three tablespoonfuls at bed-time, with a little water.

The above is given in cases of idiopathic epilepsy in which patients derive no benefit, or have ceased to have any, from the bromide or iodide of potassium alone or combined, or of the bromide of ammonium alone.

When the patient's pulse is weak, substitute for the bicarbonate of potash in the above formula the sesqui-carbonate of ammonia, and for the six ounces of infusion of columba, an ounce and a half of the tincture of that medicine with four ounces and a half of distilled water.

Dr. BROWN-SÉQUARD gives the following very important rules relative to the treatment of epilepsy by the bromide of potassium and ammonium, employed together or separately.

1. That the occurrence during the day of the sleepiness caused by these remedies can be avoided by giving relatively small doses in the day time and a much larger dose late in the evening.

2. That the quantity of these medicines to be taken each day, must be large enough to produce an evident though not complete anæsthesia of the fauces and upper parts of the pharynx and larynx; that daily quantity being from 45 to 80 grs. of the bromide of potassium and from 28 to 45 grs. of the bromide of ammonium, when only one of these salts is

employed, and a smaller quantity of each, but especially of the second, when they are given together.

3. That an acne-like eruption on the face, neck, shoulders, etc., should be produced, and it is most important to increase the dose when there is no eruption, and also when the eruption is disappearing, unless the dose already given in the twenty-four hours is so large that any increase of it causes great sleepiness in the daytime, a decided lack of will and of mental activity, dullness of the senses, drooping of the head, considerable weakness of the body, and a somewhat tottering gait.

4. That it is never safe for a patient taking either of the bromides or both, and receiving benefit therefrom, to be even only one day without his medicine, so long as he has not been at least fifteen or sixteen months quite free from attacks.

5. That the debilitating effect of the bromides in patients already weak, as are most epileptics, ought to be prevented or lessened by the use of strychnia, arsenic, the oxide of silver, ammonia, or cod-liver oil, cold douches or shower baths, and, of course, wine and a most nourishing diet. In making use of strychnia or arsenic it must be kept in mind that not only the bad influences of the bromides, but also their favorable influence against epilepsy can be diminished by these powerful agents (especially strychnia), and that it is therefore necessary, when these agents are used, to increase the dose of the bromides.

6. That iron and quinine—which are generally injurious to epileptics, except in cases in which the nervous affection is caused, or at least aggravated, by chlorosis, anæmia, or malarial cachexia—are more particularly injurious in cases in which the bromides are taken.

7. That a gentle purge every five or six weeks usually gives

a new impulse to the usefulness of the bromides against epilepsy.

64. R. Morphiæ sulphatis,   gr. ¼.
     Atropiæ sulphatis,   gr. 1-60.   M.
For one injection, in a few minims of distilled water.

Our author has succeeded in curing a case of epilepsy by
the use of this injection.

In a case of a gentleman who had pretty regular weekly
attacks of epilepsy, Dr. BROWN-SÉQUARD employed chloroform
by inhalation, almost without interruption for two or three
days successively, with the object of preventing the expected
fit or fits. It was of the greatest importance in that case to
prevent a fit, as the patient, in a preceding attack, had frac-
tured and dislocated one of his arms. The inhalation of chloroform saved him from the expected attack, and the callus
had time to be formed before he had another fit.

GEORGE JOHNSON, M. D., F. R. C. P., *Physician to King's
College Hospital, London*, speaks in high terms of chloroform
in connection with bromide of potassium in this affection. He
thinks that the action of chloroform inhalation in warding off
a threatened fit and in cutting short a violent and prolonged
paroxysm, is as uniform and certain as the action of anæmia
in exciting convulsions.

THOMAS HAWKES TANNER, M. D., F. L. S., etc., London,
has used the vapor of chloroform and believes that the fits have
diminished both in severity and numbers from its employment.

J. PHILPOT WEBB, M. D., *of Nevada City, California,
Licentiate of the Royal College of Physicians of Edinburgh, etc.,*
has recently reported a case of epileptiform convulsions arrested by chloroform inhalation in a boy aged fifteen.

MECHANICAL AND PHYSICAL MEANS.

Dr. Brown-Séquard has found : 1. That it is not necessary to apply an irritation (by a ligature, pinching, etc.,) on the very limb from which an aura seems to start, as the same means applied elsewhere may succeed ; but the chance of success is much greater by the former than by the latter way.

2. That a constant or a frequent irritation (by a blister, an issue, a seton, the actual cautery, etc.,) on the place from which an aura seems to start, may not only prevent fits, but, by some change of nutrition locally, (if the aura is really of peripheric origin,) and in the nervous centres, may reduce or even destroy altogether the tendency to fits, and lead to a complete cure.

3. That as a circular ligature may procure a temporary good effect, so a narrow *circular blister* applied all around a limb, a toe, or a finger, or a circular cauterization with a white-hot iron, may cure epilepsy in cases with a distinct aura.

4. That even in cases in which there is no aura felt or unfelt, ligatures, pinching, and other means of irritation may prevent the occurring of expected fits.

When an attack of epilepsy is followed by a comatose state, or even a sleep with heavy breathing, it is of the greatest importance to place the head of the patient in such a position that the tongue, which is then paralyzed, will not fall on the larynx and cover its aperture.

T. S. CLOUSTON, M. D., EDINBURGH.

From extensive and very elaborately conducted experiments, to determine the precise effect of bromide of potassium in epilepsy, and its proper dose, Dr. Clouston found that the diminution of the fits, and all the other good effects of the medicine reached their maximum in adults, at thirty-grain doses ter die ; while ill effects were manifested when thirty-five grain doses ter die were reached.

J. WARING-CURRAN, L. K. & Q. C. P. I., ETC.

65. R. Zinci oxidi, gr. ijss.
Ext. glycyrrhizæ, q. s. M.
For one pill. One or two ter die.

This, together with the bromide of potassium in mixture, forms a method of treatment not to be equaled in epilepsy, when assisted by the occasional application of Chapman's spinal ice bag. Neither remedy succeeds so well alone ; the one is essential to the other.

J. M. DA COSTA, M. D., PHILADELPHIA.

66. R. Zinci valerianatis, gr. iij.
Extracti belladonnæ, gr. ½.
Pulveris digitalis, gr. i. M.
For one pill.

To be taken three times a day in cases of epilepsy, associated with irregularity of the heart.

PROF. WM. A. HAMMOND, M. D., ETC., NEW YORK.

In regard to the dose of bromide of potassium in epilepsy, Dr. HAMMOND states that the symptoms due to large doses of the bromide may be enumerated as follows, in the usual order of their occurrence ; 1. Contraction of the pupils ; 2. Drowsiness ; 3. Weakness of the arms and legs ; 4. Depression of mind ; 5. Failure of memory ; 6. Delusions. The first three of these are the usual accompaniments of a dose of the medicine capable of producing any influence over epilepsy. In adults they never follow less doses than ten grains. Doses of five grains produce no effect.

HOSPITAL FOR DISEASES OF THE CHEST, LONDON.

67. R. Potassii bromidi,   gr. x.
         Tincturæ conii,   ℳ xxx.
         Tinc. valerianæ ammoniatæ,   ℳ x.
         Aquæ camphoræ,   f.ʒj.   M.
     For one dose ; ter die.

HOSPITAL OF UNIVERSITY COLLEGE, LONDON.

68. R. Potassii bromidi,   gr. x.
         Spiritus chloroformi,   gtt. xviij.
         Infusi quassiæ,   f.ʒj.   M.
     For one dose ; ter die.

DR. MARSHALL HALL.

69. R. Strychniæ acetatis,   gr. j.
         Acidi acetici,   ℳ xx
         Alcoholis,   f.ʒij.
         Aquæ destillatæ,   f.ʒvj.   M.
     Ten drops (gr. 1–50) to be taken in water, ter die.

WALTER TYRRELL, M. R. C. S., states that he has watched
the effects of strychnia upon various forms of epilepsy and has
no hesitation in affirming that in a large majority of cases its
effects are most beneficial.   He found but three cases in which
it produced no favorable result, and no case in which it produced an unfavorable effect.   He gives a medium quantity as
a dose, for a lengthened period, rather than carry the dose too
high at first.   The best results are obtained from gr. 1-10 to
gr. 1-18 twice a day in solution, the system appearing to
regain its nervous strength under the continued use of the
medicine.

J. SPENCE RAMSKILL, M. D., LONDON, PHYSICIAN TO THE HOS-
PITAL FOR THE PARALYZED AND EPILEPTIC.

70. R. Bruciæ,   gr. iv.
         Alcoholis,   f.ʒij.
         Aquæ destillatæ,   f.ʒvj.   M.
     Ten minims to be taken diluted with water twice daily ;
        every third day an addition of five minims should be
        made to the dose until from a third to a half grain is reached.
     In the treatment of stomachal epilepsy.

If any stiffness of the jaws, or other toxic symptoms appear, the dose is to be diminished five minims, and continued until any new objectionable symptom is manifested, then it is again lessened. No benefit will be derived until a full dose is reached; often the reverse effect. As a rule, patients will take twice as much brucia as strychnia without any necessity for diminishing the dose. After the continuous administration of brucia for a month, it is well to suspend its use for some days, and then again resume it. Great satisfaction will be obtained by giving the bromide of potassium in large doses at bed-time, and at the same time ordering brucia twice daily, thus insuring the sedative influence of the bromide and the tonic effect of the brucia on the whole nervous system.*

CHOREA.

WILLIAM AITKEN, M. D., EDINBURGH.

The indications of cure are:—1. To remove, if possible, all morbid states of the body which may tend to aggravate the disease, such as constipation, anæmia, amenorrhœa, worms; 2. By well regulated purgative medicines to subdue any cerebral congestion; 3. To sustain the strength and improve the vigor of the nervous system by tonic and stimulant medicines, by food and by the cold bath.†

71. R. Camphoræ, ℈v.
    Syrupi, q. s. M.
    Divide into xx pills. One three times a day. Useful after discharges have become healthy by the action of the purgatives.

72. R. Spiritus ætheris nitrosi, f.℥j.
    Mis'uræ camphoræ, f.℥iij. M.
    A tablespoonful three times a day.

*London Lancet, Jan. 16, 1869, p. 75.
†The Science and Practice of Medicine, Am. Ed., 1868; vol. ii, p. 337.

32   MODERN THERAPEUTICS.

Many young women, who attribute the attack to fright, get well under this treatment.

THOMAS KING CHAMBERS, M. D., ETC., LONDON.

73. ℞. Liq. potassæ arsenitis,   ♏ v.
Ter die; to be increased to   ♏ xvj.
Also cod liver oil and iron, if indicated by the general condition.

Injudicious management of patients afflicted with chorea frequently protracts the case. One of the most common forms of injudicious management is the fixing of the attention of patients upon their infirmity, by telling them how bad they are, offering unnecessary help, etc. They should be encouraged to make every exertion to direct the movements of the limbs; as by slow walking to music, carrying trays and crockery, and other things that demand care. In order that their attention may be withdrawn from their deficiencies, looking glasses and the distressing sight of other choreics should be avoided. They should be got away from home as soon as possible. Sent under the care of a judicious person to the seaside, or anywhere else for an excuse, children often recover rapidly; whereas, had they remained at home, they would have continually relapsed.

J. M. DA COSTA, M. D., PHILADELPHIA.

74. ℞. Zinci valerianatis,   ℈ij.
Cinchoniæ sulphatis,   ℈j.   M.
For xx pills; one ter die.

Frequently a partial loss of power in children coincides with the setting in of chorea—a sign of debility of the nervous centres, particularly of the spinal cord, and to be treated by tonics conjoined with antispasmodics as in the above recipe.

75. ℞. Cupri ammoniati,   gr. |.
      In pill ter die; to be gradual-
      ly increased to   gr. j.
76. ℞. Ext. cimicifugæ fluidi,   gtt. xx
      For one dose, three times a day.

THOMAS HILLIER, M. D., LOND., F. R. C. P., ETC.

77. ℞. Liq. potassæ arsenitis,   ℥ij.
      Potassæ bicarbonatis,   gr. iij.
      Potassii iodidi,   gr. ij
      Aquæ camphoræ,   f.℥ss   M.
For one dose, ter die, to children aged five, for aggravated
chorea, attended with severe pains in the limbs, and rheumatic persistent swellings.

Arsenic in full doses is a valuable remedy in a fair proportion of cases, but in some instances it entirely fails. Iodide
of potassium is useful when the patient is subject to chronic
rheumatism.

Occasionally purgatives and tonics, especially iron, are attended with much success. Strychnia, so highly recommended
by Trousseau, seems, to our author, to be highly injurious in
the acuter stages of the disease ; in the more chronic form, and
where there is a tendency to paralysis, it is of service. Iron
and strychnia may be combined thus:

78. ℞ Strychniæ,   gr. 1-32.
      Vini ferri,   f.℥ij.   M.
For one dose, ter die, to a child ten years of age.

Narcotics such as opium, belladonna, cannabis indica, or
conium, are of little or no use. Anti-spasmodics, such as
valerian, and assafœtida, are also useless.

Our author has seen good results from the employment of
baths of sulphuret of potassium.

79. ℞. Potassii sulphureti,   ℥iv.
      Aquæ, (90° F.)   C.xxx.
      For a bath ; the patient to remain in it for an hour
daily.

This bath is also recommended by Dr. FELIX VON NIE-
MEYER, when there is anæmia.

Gymnastic exercises, shampooing, and passive movements
are of service. As many muscles as possible should be exercised, without fatiguing any of them. Shower-baths are useful in the latter stages, when the patient is not timid, or too
much excited by them.

Dr. NIEMEYER quotes *Benedikt*, who declares that out of
more than twenty cases of chorea, treated by him by the
constant galvanic current, not one has failed to recover. The
current which he employs is just strong enough for the patient to feel it distinctly, and he applies it along the spine, the
patient standing erect. Painful currents aggravate the symptoms.

DR. J. W. OGLE.

80. R. Pulveris physostigmatis,   ℥j.
       Alcoholis,   f.℥j.   M.

Begin with twenty minims, ter die, and increase, by ten
    minims a dose, to f.℥j. Our author reports several
    cases treated in this manner successfully.

DR. H. ROYER, FRANCE.

81. R. Sodæ arseniatis,   gr. j.
       Syrupi acaciæ,   f.℥iv.   M.

Dessertspoonful three times a day.*

THOMAS HAWKES TANNER, M. D., F. L. S., ETC.

82. R. Zinci phosphatis,   gr. xx-xl.
       Acidi phosphorici diluti,
       Tincturæ ferri chloridi,   aa f.℥jss.
       Aquæ menthæ pip.,   q. s. ad f ℥vj.   M.

Two tablespoonfuls ter die.

The only plan to be followed, in treating chorea, consists
in regulating the bowels, subduing irritation, and strengthen-

*Formulaire Raisonné des medicaments Nouveaux, etc., Par. O. Reveil, Deuxieme
edition, p. 347.

ing the system. For the first purpose, calomel and jalap, or, when worms are suspected, oil of turpentine may be employed. A combination of tonics or anti-spasmodics, with purgatives, is often serviceable. The two great remedies are the cold shower or douche bath, and iron. The former should be employed every morning, on the patient's rising. Cod-liver oil is generally useful, administered with tonics. Mental excitement should be guarded against, and nutritious food and exercise in the fresh air insisted upon.

DR. JAMES TURNBULL, OF LIVERPOOL.

83. R.   Anilinæ sulphatis,   ℥ss.
     Divide into xx powders, one to be taken three times a day.*

The sulphate of aniline has also been given in as large doses as three grains every third hour. It is a white powder, easily taken.†

HYSTERIA.

WILLIAM AITKEN, M. D., EDIN.

The following directions are given by our author as to what may be done during a fit of hysteria. Everything tight about the patient's person should be loosened. The window should be opened and the cold air allowed to blow over her. The horizontal posture on a bed or the floor should be secured. This being done, many modes of further proceeding may be followed. Bleeding is in all cases of doubtful efficacy. When the jaw is locked the following enema (recommended by Dr. Wood) may be used

---

* Aitkin's Science and Practice of Medicine, Am. Ed. 1869, vol. ii. p. 333.
† *Half-Yearly Compendium of Medical Science. Jan., 1869, p. 71.*

84. ℞. Assafœtidæ,　　　　　　　ʒij.
　　　　Aquæ.　　　　　　　　　　f.Oss.　　　M.
To be beaten up with the yolk of an egg, or, what is still
　better,

85. ℞. Olei terebinthinæ,　　　　　f.ʒss.
To be mixed with the yolk of an egg, and then added to
　half a pint of water.

Another remedy is to fill the mouth with salt. But that
which supersedes all others and is unquestionably the best,
is a good drenching with cold water. If the patient lie on
the bed the head should be drawn over its side, and a large
quantity of water poured on it from a considerable height out
of a pail, jug or other large vessel, and directly over the
mouth and nose of the patient, so as to stop her breathing
and compel her to open her mouth. This practice is generally introduced into hospitals, and until it was adopted, it
was not unusual to see three or four patients in hysteria in
the same ward and at the same time. Under this practice,
however, an hysterical case is rare, and the fit seldom occurs twice in the same person and never becomes epidemic.*

THOMAS KING CHAMBERS, M. D., ETC., LONDON.

86. ℞. Acidi muriatici diluti,　　　f ʒjss.
　　　　Aquæ calefactæ,　　(95° F.)　C.xxx.　M.
For a bath. This tonic warm bath is to be used once a day,
　in order to prepare the patient for a *shower-bath* twice
　a day.

Shower-baths in hysterical cases are highly recommended
by Dr. C. The making up the mind to the shock of a cold
shower-bath is a capital exercise of the will. Such baths
have also a good influence by arterializing the cutaneous circulation, driving the venous blood home to the heart and
lungs.

---

*Science and Practice of Medicine. Am. Ed. 1868, vol. ii. p. 345.

Our author rings the changes upon the following prescriptions in the treatment of this disease:

87. ℞. Pilulæ assafœtidæ, No. xxx.
Three to be taken ter die.

88. ℞. Spiritûs ammoniæ fœtidæ, f.℥iij.
A teaspoonful in water three times a day.

89. ℞. Tincturæ castorei ammoniatæ,
Aquæ fœniculi, aa f.℥ij.
A dessertspoonful in water ter die.

90. ℞. Pil. galbani comp., No xxx.
Two ter die

91 ℞. Zinci valerianatis, ℨj.
Syrupi, q. s.
Divide into xx pills. One to be taken three times a day.

SIR CHARLES LOCOCK.

92. ℞. Potassii bromidi, ℨijss.
Aquæ cinnamomi, f.℥iv. M.
A dessertspoonful thrice daily.

In hysterical epilepsy with disordered uterine functions, the
treatment being prolonged. It was this use of bromide of
potassium suggested by our author which led to its introduction as a remedy in forms of epilepsy other than the hysterical.

DR. FELIX VON NIEMEYER, PROFESSOR OF PATHOLOGY AND
THERAPEUTICS, ETC., UNIVERSITY OF TUBINGEN.

93. ℞. Aurii et sodii chloridi, gr. v.
Gummi tragacanthi, ℨj.
Sacchari albi, q. s. M.

Divide into xl pills. Order at first one of these pills to be
taken an hour after dinner, and another an hour after
supper. Afterwards order two pills to be taken at these
hours and gradually increase the dose up to eight pills
daily.

Our author speaks of this preparation as a nervine of great
efficacy in hysteria. He has made use of it with signal effect

in many cases where there was no indication for the local treatment of uterine disease, or else, where the hysteric symptoms persisted although the local uterine affection had been cured.*

THOMAS HAWKES TANNER, M. D., F. L. S., ETC., LOND.

94. R. Tincturæ assafœtidæ,   f.ʒij.
     Ammoniæ carbonatis,   ʒj.
     Aquæ camphoræ,   q. s. ad. f.ʒiv.   M.
One or two tablespoonfuls occasionally when feeling languid or hysterical.†

95. R. Tincturæ assafœtidæ,   f ʒij.
     Spiritûs ammoniæ aromatici,   f.ʒiij.
     Tincturæ chirettæ,   f.ʒvij.
Sixty drops in a wineglassful of water every two or three hours until the paroxysms cease.

96. R. Tinc. valerianæ ammoniatæ,   f.ʒijss.
     Infusi valerianæ,   f.ʒiv.
Two tablespoonfuls to be taken occasionally.

97. R. Ferri phosphatis,   ʒij.
     Acidi phosphorici diluti,   f.ʒjss.
     Syrupi aurantii corticis,   f.ʒj.
     Syrupi acaciæ,   f.ʒiij.   M.
A tablespoonful, largely diluted, three times a day.

98. R. Phosphori,   gr.j.
     Olei amygdalæ dulcis,   f.ʒiij.   M.
One teaspoonful in a wineglassful of barley water three times a day.

The patient's diet should be regulated. She should have nourishing food, and often a moderate quantity of wine or beer. Hot rooms and evening parties are to be proscribed, and stays ought not to be worn. Healthy mental occupation should be afforded.

---

*Text Book of Practical Medicine. Am. Ed. 1868, vol. ii., p. 385.
†Practice of Medicine, Am. Ed. 1865, p. 296.

EDWARD JOHN TILT, M. D., R. C. P., ETC. LONDON.

90. R. Tincturæ castorei, f.ℨiij.
Spiritus lavandueal composit', f.ℨvj.
Misturæ camphoræ, q. s. ad f.ℨvj. M.

A tablespoonful two or three times a day, when cerebral symptoms and hysterical phenomena are marked.

The therapeutical indications in the treatment of hysteria are : 1st. To blunt the sensitiveness of the nervous system by sedatives and anti-spasmodics, and to strengthen it by metallic and other tonics, and by hygiene. 2d. To cure all diseases of the sexual organs, and save the nervous system from visceral irritation, by good hygiene at menstrual periods, or by marriage, when the sexual organs crave for their legitimate satisfaction.*

WILLIAM FENWICK, M. D., GLASGOW.

100. R. Pulveris physostigmatis,
" rhei, aa ℨj. M. ,

Divide into xx powders. One to betaken every four hours during the day, also an occasional dose at night, making the average quantity of fifteen grains of each in twenty-four hours.

Under the influence of this combination, Dr. F. has seen none of the depressing effects which the bean produces by itself. He reports the improvement under this treatment as marked.†

TETANUS.

HYPODERMIC INJECTION.

DR. E. FICK, OF REVAL.

101. R. Morphiæ sulphatis, gr. ¼.
Aquæ, ♏ vj. M.

For one injection, between the shoulder blades near the spine ‡

*Hand-Book of Uterine Therapeutics, Am. Ed. 1869, p. 70.
†Glasgow Medical Journal, May, 1869, p. 300.
‡Half-Yearly Compendium of Medical Science, Jan., 1869, p. 3.

G. OLLIVER, M. D., LONDON, M. R. C. S., ETC.

102. R. Atropiæ,   gr. 1-60.

In the form of a granule, one every three hours ; and linimentum belladonnæ to be rubbed over the spine and rigid
muscles every six hours.

A successful case of treatment by this method is reported
in the *British Medical Journal* for August 22d, 1868. The
patient was kept under the influence of atropia for three weeks.
He then quickly and completely recovered his usual health
under iron and quinine.*

C. V. RIDENT, M. R. C. S.

103. R. Extracti physostigmatis, gr. viij.
         Alcoholis,   f.ʒj.   M.

The extract of calabar bean to be well rubbed down in the
alcohol. The dose is ten minims (—gr. 1-6 of the extract).

Our author records a case† of traumatic tetanus in which
this dose, ten minims of the tincture, was given every hour. It
appeared to retard the progress of the case and to ameliorate
the severity of the symptoms, but never fully to control them.

The following directions are given in the *British Pharmacopœia* for making the

EXTRACT OF CALABAR BEAN.

104. R. Calabar Bean, in coarse powder, 1 pound.
         Rectified spirit,   4 pints.

Macerate the bean for forty-eight hours with one pint of the
spirit in a close vessel, agitating occasionally, then transfer to
a percolator, and when the fluid ceases to pass add the re-
mainder of the spirit so that it may slowly percolate through
the powder. Subject the residue of the bean to pressure, adding
the pressed liquid to the product of the percolation ; filter,

---

*Half-Yearly Compendium of Medical Science, July, 1868, p. 77.
†Lancet, Oct. 31, 1868, p. 567.

distill off most of the spirit, and evaporate what is left in the retort by a water-bath to the consistence of a soft extract.

Dose—1-16 to 1-4 grain.

PROGRESSIVE LOCOMOTOR ATAXIA.

DR. W. LAMBERT, M. B., AMHERSTBURG, ONTARIO, CANADA.

105. R. Acidi phosphorici diluti,   f.ʒvj.
          Syrupi simplicis,   f.ʒiij.

A teaspoonful in water ter die, gradually increased to a dessertspoonful, together with the application of electricity.*

EBEN. WATSON, M. D., PROFESSOR OF PHYSIOLOGY IN ANDERSON'S UNIVERSITY AND SURGEON TO THE ROYAL INFIRMARY, GLASGOW.

106. R. Tincturæ physostigmatis,   ♏ v-xxv.
          For one dose, to be given every half hour.†

Five minims of the tincture equals one-half grain of the extract of calabar bean.

NEURALGIA.

WILLIAM AITKEN, M. D., EDIN.

When the neuralgia is superficial, compresses steeped in the solution of atropia have a good effect.

107. R. Atropiæ sulphatis,   gr. v.
          Aquæ destillatæ,   f.ʒiij.

Renew the compresses several times in twenty-four hours. Continue them for at least an hour each time, and cover them with oil skin to prevent evaporation. This is the formula recommended by Trousseau.‡

---

* N. Y. Medical Journal, February, 1869, p. 482.
† *The Practitioner*, Sept. 1869, p. 146.
‡ Science and Practice of Medicine, Am. Ed. 1868, vol. ii. p. 524.

BROWN-SÉQUARD.

108. R. Extracti belladonnæ,   gr. 1-6.
      "   stramonii,   gr. 1-5.
      "   cannabis indicæ,   gr. 1-4.
      "   aconiti,   gr. 1-3.
      "   opii,   gr. 1-2.
      "   hyoscyami,   gr. 2-3.
      "   conii,   gr. j.
      Pulveris glycyrrhizæ,   q. s.
For one pill.

According to circumstances *Brown-Séquard* gives, without producing any great constitutional disturbance, three, four, and even five pills in a day, and sometimes in about eight or ten hours, for the relief of neuralgic or other pains. There must be, therefore, some influence exerted by some of these substances upon the others, diminishing their bad and not their good effects.

HYPODERMIC INJECTION.

109. R. Morphiæ sulphatis,   gr. ½-⅓.
          Atropiæ sulphatis,   gr. 1-25.
For one injection, in gtt. xx of distilled water.

The doses BROWN-SÉQUARD at first employed were gr. 1-2 of the sulphate of morphia to gr. 1-60 of the sulphate of atropia. He now employs those given above, the antagonistic effects of morphia and atropia on the brain rendering it possible, while securing the good effects against pain of the two remedies, to use safely, or at least without great or lasting cerebral or cardiac disturbance, large doses of these narcotics.

In this connection, we may give the doses employed by other authorities in administering morphia and atropia subcutaneously.

Dr. CHARLES HUNTER, of London, gives, as a rule, never to use, in the first injection, in any case, more than one-half the stomachic dose of these alkaloids for males, and not more than a third for females.

Dr. RUPPANER, of New York, places the minimum dose of morphia at gr. 1-8, the maximum, gr. 3-4; the minimum dose of sulphate of atropia, gr. 1-60, the maximum, gr. 1-30.

Dr. TILT, of London, states that the initial hypodermic dose of morphia (used alone) for a woman, should not exceed gr. 1-6 (the acetate being the salt he prefers); and that of atropia should not exceed (used alone) gr. 1-100. In combination, gr. 1-6 of the sulphate of morphia with gr. 1-30 of the sulphate of atropia.

J. M. DA COSTA, M. D., PHILADELPHIA.

110. ℞. Aconitiæ,   gr ¼s.
         Veratriæ,   gr. xv.
         Glycerinæ,   f.ʒij.
         Cerati adipis,   ʒvj.   M.
To be rubbed over the painful parts, care being taken to see that there is no abrasion of the skin.

(Prof. GROSS sometimes employs veratria ointment in cases of neuralgia, of the strength of ʒj. to ʒj.)

111. ℞. Potassii bromidi,   ʒss.
         Succi conii,   f.ʒx.
         Aquæ cinnamomi,   q. s. ad f.ʒiij. M.
A dessertspoonful to be taken three times a day.

Useful in epileptiform neuralgia.

112. ℞. Strychniæ sulphatis,   gr. ¼.
         Quiniæ sulphatis,   gr. xv.
         Cinchonæ sulphatis,   ʒj.
         Pulveris rhei,   ʒss.
         Ext. gentianæ,   q. s.   M.
Divide into xxx pills. One to be taken four times a day, in intercostal neuralgia of malarial origin, associated with constipation.

M. DUMAS, MONTPELLIER, FRANCE.

113. ℞. Castorei,   gr. xxx.
         Camphoræ,   gr. xv.
         Pulveris opii,   gr. viij.
         Confectionis rosæ,   q. s.
To be divided into xv pills, and used for nervous affections of the abdominal organs in women.

DR. GRAY, FRANCE.

114. R. Tincturæ aconiti,
     Chloroformi,   aa f.ʒijss.
     Adipis,   ʒx. M.
After applying the ointment to the affected part, the place
 is covered with cotton.*

PROF. S. D. GROSS, M. D., PHILADELPHIA.

115. R. Quiniæ sulphatis,   Əij.
     Morphiæ sulphatis,   gr. j,
     Strychniæ,   gr.2–3
     Acidi arseniosi,   gr. j.
     Extracti aconiti,   gr. x. M.
Divide into xx pills. One to be taken three or four times
 a day. Add to the recipe, ferri sulphatis, Əij., if the system be anæmic.

This pill is useful in a great variety of cases of neuralgia.
Its effects should, of course, be carefully watched.

GUY'S HOSPITAL, LONDON.

116. R. Liq. plumbi subacetatis,
     Tincturæ opii,
     Mellis,   aa f ʒij.
     Confectionis rosæ,   ʒj. M.
 Fiat linimentum.

This is known in the pharmacopœia of the hospital as linimentum plumbi opiatum, and is much used.

LONDON HOSPITAL.

117. R. Tincturæ aconiti,
     Linimenti saponis,   aa f.ʒj. M.
 To be used as an anodyne liniment.

PROF. WILLIAM A. HAMMOND, M. D., ETC., N. Y.

118. R. Extracti belladonnæ,   gr. v.
Divide into xx pills. One ter die, the dose to be increased
 as necessary.

The use of belladonna is chiefly to change the habits of the
system. This drug, although at one time much overlauded, is

---

*Half-Yearly Compendium of Medical Science, Jan., 1863, p. 49.

very efficient in the treatment of neuralgia. Our author has not used atropia often, as the dose is difficult to graduate.

Hypodermic injections of morphia may be used during the paroxysms of pain. In their use avoid the face; a good point is the inside of the arm.

> 119. R. Tincturæ aconiti, f.ʒss.
> Rub with a rag upon the painful part until a sense of prick-
> ling is felt.

This is next in value to the subcutaneous use of morphia. The action is often very powerful. Dr. H. once caused temporary paralysis of the arm in a lady by the too free application of the tincture.

Chloroform may be used externally, internally, or by inhalation not carried to insensibility. Repetitions of inhalation may break up the paroxysm.

Hypophosphites are useful; may be given in doses of from ten to twenty grains. They act by setting free phosphorous in the stomach.*

Galvanism.—The direct galvanic current may be successfully used both for the relief of the paroxysm and for breaking up the habit of the disease. For the first apply the poles so that the position is near the seat of the greatest pain, and pass the current continuously for several minutes. To change the habit of the system, apply one pole to the nape of the neck, and the other over the course of each sympathetic nerve, moving it along the neck. DUCHENNE says, that neuralgia can generally be relieved by Faradization. Dr. H. has not been so uniformly successful, but often effects a cure by the application mentioned.

---

*Half-Yearly Compendium of Medical Science, July, 1868, p. 67.

The use of tea generally aggravates neuralgia, while coffee, on the contrary, does not, but if strong is often of service.*

THOMAS HAWKES TANNER, M. D., F. L. S., ETC., LOND.

120. R. Quiniæ sulphatis,   gr. xxiv.
          Extracti balladonnæ,   gr. iv.
          Camphoræ,   gr. xxx.
          Confectionis rosæ,   q. s.
Divide into twelve pills. One to be taken two or three times a day, in cases of neuralgia in which the attacks are periodic.†

121. R. Quiniæ sulphatis,   ℈j.
          Liquoris arsenici chloridi,   f℥iij—iv
          Acidi sulphurici aromatici,   f℥ij.
          Syrupi zingiberis,   q. s. ad. f.℥iij. M.
One teaspoonful in two tablespoonfuls of water directly after breakfast, dinner and tea, in severe neuralgia.

122. R. Ammoniæ muriatis,   ℥iij.
          Aquæ,   f.℥iij   M.
A tablespoonful in water every hour, while the paroxysm of pain is on. If after the fourth dose there be no diminution of pain, it will be useless to persevere. As soon as the pain is relieved the dose may be reduced to a dessert-spoonful three times a day.

RECTAL SUPPOSITORIES.

EDWARD JOHN TILT, M. D., ETC., LONDON.

124. R. Extracti hyoscyami,   ℥j.
          "   belladonnæ,   gr. v.
          Butyri cocoæ,   q. s.   M.
Make into xx suppositories, round, in pill form. One to be introduced at night. This is the suppository Dr. Tilt most frequently prescribes, for it relieves pain without constipating.

123. R. Extracti opii,   ℈j.
          "   belladonnæ,   gr x.
          Butyri cocoæ,   q. s.   M.
Make into xx round suppositories. One to be well introduced into the bowel at night.

---

*Half-Yearly Compendium of Medical Science, July 1858, p. 67.
Half-Yearly Compendium of Medical Science, July. 1869, p. 65.
† Practice of Medicine. Am. Ed. 1866. P. 318.

VAGINAL SUPPOSITORY.

125. R. Morphiæ sulphatis, gr. iij.
Butyri cocoæ, q. s. M.
Divide into vj. suppositories. One to be used at night.

ENEMA.

126. R. Battley's solution of opium, f.ʒj.
Tincturæ hyoscyami, f.ʒj.
Aquæ, f.ʒiij. M.
One tablespoonful of this, or double the quantity, to be added to a little warm milk and injected.

LINIMENTS.

127. R. Morphiæ sulphatis, gr. viij.
Atropiæ sulphatis, gtt. iv.
Olei rosæ, gtt. ij.
Alcoholis, f.ʒss.
Olei olivæ, q. s. ad f.ʒiv. M.

This liniment should be shaken before it is used. The sulphate of atropia is preferable to atropia, because it is more soluble, and oil is much better than glycerine as a constituent of liniments.

Another excellent and elegant and sedative liniment is the following:

128. R. Atropiæ sulphatis, gr. viij.
Morphiæ sulphatis, gr. gr. xvj.
Aconitiæ, gr. ij.
Acidi sulphurici diluti, ♏, v.
Alcoholis, f.ʒss.
Olei olivæ, q. s. ad. f.ʒiv. M.

Or, if a stimulant effect is also desired,

129. R. Chloroformi, f.ʒss.
Spiritûs terebinthinæ, f.ʒj.
Camphoræ, ʒij.
Olei lavandulæ,, ♏ xx.
Olei olivæ, q. s. ad f.ʒvj. M.

The first four ingredients should be mixed before adding the oil, and the liniment should be well shaken before it is applied.

OINTMENT.

130. R.   Atropiæ sulphatis,   gr. ij.
            vel
           Morphiæ sulphatis,   gr. x.
           Glycerinæ,   f.ℨss.
           Olei neroli,   gtt. iv.
           Unguenti glycerinæ,   ℨj.
      To be rubbed into the skin twice a day.

Dr. Aug. Waller has found that certain substances, such as atropia, strychnia, morphia, and the tincture of aconite, when mixed with chloroform and applied to the skin, are absorbed rapidly; but if alcohol is used instead of chloroform, absorption is delayed or altogether prevented. The ability of introducing rapidly into the blood these active narcotics, without the use of the hypodermic syringe, will be of much importance in those cases in which even a slight puncture of that instrument is dreaded. It will also be of value in those instances in which it is necessary or advisable to keep up the effect for a long time, as in hydrophobia, chronic neuralgia, etc.

FACIAL NEURALGIA.

FRANCIS E. ANSTIE, M. D., F. R. C. P., ETC., OF LOND.

In the treatment of that terrible kind of facial neuralgia, to which Trousseau gave the name of *epileptiform*, Dr. A. recommends the following plan in the earlier stages of the malady:

1. *Counter-irritation* applied, not to the branches of the fifth, but to those of the occipital nerve, at the nape of the neck. A blister in the former situation is often as hurtful as useful; in the latter it is sometimes strikingly effective in gaining a short respite.

2. *Nutritive Tonics.*—The assiduous use of cod-liver oil, or of some fatty substitute for it, should be insisted on from the first, and is of the highest consequence.

3. *Subcutaneous injection* of morphia, or of atropia, accord-

ing to circumstances. Commence with the use of one-sixth of a grain of morphia twice daily, increasing this, if necessary, to one-fourth and one-half a grain, and in rare cases to one grain. If this produces, along with the other measures, a notable remission of the pain, it should be cautiously and steadily decreased, as circumstances may admit. In cases where morphia fails, atropia may be tried in doses commencing at one-sixth of a grain. The injection of a less quantity than this would probably be useless in severe tic.*

J. M. DA COSTA, M. D., PHILADELPHIA.

131. R. Liquoris potassæ arsenitis, gtt. v.
Syrupi rhei aromatici, f.ʒj. M.
For one dose, ter die, after meals.

At night apply a hot salt bag to the back of the neck and order the following pill:

132. R. Extracti belladonnæ, gr. ¼.
Extracti hyoscyami,
Extracti colocynth. comp.,
Pulv. zingiberis, aa gr. j. M.

ASHLEY N. DENTON, M. D., SUTHERLAND SPRINGS, TEXAS.

133. R. Olei camphoræ, f.ʒij.
Pulveris opii, ϶j.
Potassæ nitratis, gr. xv. M.
Mix well in a mortar, and apply to a denuded surface. Useful in relieving the intense suffering from facial neuralgia.†

Dr. D. also recommends chloroform, as advised by Dr. Tanner, as one of the most useful palliatives in his hands.

---

* *Half-Yearly Compendium of Medical Science.* July, 1869, p. 186.
† St. Louis *Medical Reporter*, June 15th, 1867, p. 264.

DR. FELIX VON NIEMEYER, PROFESSOR OF PATHOLOGY AND THERAPEUTICS, UNIVERSITY OF TUBINGEN.

134.   R.   Extracti hyoscyami,
        Zinci oxidi,   aa ℈ij.   M.

Divide into xl pills. Begin with one pill, morning and evening, and increase to twenty or thirty of them daily. These, known as Meglin's pills, have a good reputation in Germany.*

THOMAS HAWKES TANNER, M. D., F. L. S., ETC. LOND.

135.   R.   Extracti belladonnæ,
        Extracti opii,   aa ℨiss.
        Glycerinæ,   f.ℨiv.
        Extracti Papaveris,   ℨiss.   M.

To be painted over the affected part. A fomentation, flannel or hot linseed poultice, is to be applied, being separated from the extracts by a sheet of tissue paper.†

136.   R.   Aconitiæ,   gr. ij.
        Alcoholis,   gtt.vj.
        Mix thoroughly and add
        Adipis,   ℨj.   M.

A small portion to be cautiously smeared over the track of the painful nerve once or twice a day; but it must not be used where there is the slightest abrasion.

EDWARD WAAKES, M. D., LOND., F. L. S., LUTON.

137.   R.   Potassæ bicarbonatis,   ℨiss.
        Extracti ergotæ fluidi,   f.ℨj.
        Infusi ergotæ,   f.ℨvj.   M.

Two tablespoonfuls every four hours, in Tic Doloreux.

HEMICRANIA.

M. BERTRAND, PARIS.

138.   R.   Veratriæ,   gr. v.
        Morphiæ sulphatis,   gr. iij.
        Adipis,   ℨj.   M.

* Text-Book of Practical Medicine, Am. Ed., 1869, vol. 2, p. 295.
† The Practice of Medicine. Am. Ed., 1866, p. 318.

The painful parts to be rubbed with this ointment frequently when the paroxysms of pain are at their height, and as often as they require. Two or three frictions suffice in the majority of cases. M. Bertrand has published a number of cases of facial neuralgia, and of neuralgic headache, in which quinine and blisters had been tried without effect, and which yielded promptly to this application.

J. M. DA COSTA, M. D., PHILADELPHIA.

139. ℞. Extracti conii fluidi.   ℳij.
        Ammoniæ muriatis,   gr. v.
        Syr. aurantii corticis,
        Aquæ,   aa f.℥ss.   M.

For one dose, ter die.

In neuralgic headache associated with plethora rather than anemia in young women. Also, a drachm of cream of tartar before breakfast, daily, or if necessary, twice a day.

GEO. H. NAPHEYS, M. D.

A small cup of strong coffee, with which the juice of a lemon is mixed, will in some cases afford marked relief in attack of hemicrania.

DR. FELIX VON NIEMEYER, PROF. OF PATHOLOGY AND THERAPEUTICS, ETC., UNIVERSITY OF TUBINGEN.

140. ℞. Caffein citratis,   gr. x.
        Syrupi,   q. s.   M.
     Divide into ten pills. One to be given every hour on the first symptoms.*

An infusion of unroasted coffee drank daily, appears sometimes to render the attacks more rare and less severe.

(For other recipes see Headache.)

SCIATICA.

J. M. DA COSTA, M. D., PHILADELPHIA.

141. ℞. Emplastri epispasticæ, 1½in.x5in.

To be applied over the affected part. Let it draw for five or six hours; poultice it, and then remove the cuticle and dress with

142. ℞. Morphiæ sulphatis,   gr. ¼.
          Pulveris marantæ,   gr. ij.   M.

For one powder. Also ten grains of Dover's powder, to be taken at night.

ELECTUARY.

DR. FELIX VON NIEMEYER, PROFESSOR OF PATHOLOGY AND THERAPEUTICS, ETC., UNIVERSITY OF TUBINGEN.

143. ℞. Olei terebinthinæ,   f.ʒj.
          Mellis,   ʒj.   M.

A tablespoonful twice daily.

ROMBERG also speaks very highly of this as a specific remedy in the treatment of sciatica.*

THOMAS HAWKES TANNER, M. D., F. L. S., ETC., LONDON.

144. ℞. Sodæ sulphatis,   Эij-iv.
          Sodæ carbonatis,   Эj.
          Sodii chloridi,   gr.xv.
          Cretæ preparatæ,   gr.x.
          Ferri carbonatis sacharatæ, gr.xv.   M.

Make a powder and direct it to be taken early in the morning in half a pint of water.

In some cases in which Dr. T. could detect no cause for the sciatica, a cure has been effected by this treatment, with the use of the hot-air bath twice a week.

EDWARD WAAKES, M. D., LOND., F. L. S., LUTON.

145. ℞. Extracti ergotæ fluidi,   f.ʒij.
          Aquæ cinnamomi,   f.ʒiij.   M.

A dessert spoonful in water, every four hours.

Tinctura ferri chloridi may be added if indicated. When ergot is likely to be useful its good effects commence immediately.

*Text Book of Practical Medicine, Am. Ed., 1869, vol. 2, p. 306.

II. DISEASES OF THE RESPIRATORY SYSTEM.

CORYZA.

J. SOLIS COHEN, M. D., PHILADELPHIA.

INHALATION.

146. R. Pulveris aluminis,   gr.v-xxx.
       Aquæ,   f.ʒj.   M.

The nebulized spray to be drawn into the nostrils three or four times a day to diminish the profuse secretion of coryza and destroy fetor when present.*

DR. LOMBARD, OF GENEVA.

147. R. Pulveris opii,
       "   benzoinii,
       Sacchari albi,   aa gr. ij.   M.

For one powder. To be used in *catarrhal neuralgia complicating coryza*, in the following manner: Heat in the fire a thin plate of iron—a shovel—and throw upon it this powder, holding the head over it, so as to breathe the fumes through both nose and mouth. To be repeated two or three times a day, or oftener. It acts like a charm, frequently giving immediate relief.†

M. LUC, SURGEON, FRENCH ARMY.

Our author recommends the inhalation of iodine vapor in this affection, effected by placing a bottle of the tincture under the nose, the hand supplying warmth enough to vaporize the iodine. The inhalations, each lasting a minute, are to be repeated every three minutes during an hour.‡

J. FORSYTH MEIGS, M. D., PHILADELPHIA.

148. R. Unguenti hydrarg. nitratis,   ʒss.
       Extracti belladonnæ,   gr. x.
       Adipis,   ʒss.   M.

---

*Inhalation: Its Therapeutics and Practice, (L. & B.) Philada., 1869. p. 102.
†Dickinson's Elements of Medicine, p. 328.
‡Inhalation: Its Therapeutics and Practice. J. Solis Cohen, M. D., p. 213.

Useful in chronic coryza of children; to be completely softened by gentle heat, and applied at night on a camel's
hair pencil, to the surface of the mucous membrane
itself, and not merely to the outside of the hardened scabs.
Injections should be employed during the day.

DR. FELIX VON NIEMEYER, PROFESSOR OF PATHOLOGY AND
THERAPEUTICS, UNIVERSITY OF TUBINGEN.

In the treatment of acute nasal catarrh, the production of
active diaphoresis is the only procedure worthy of confidence.
A Russian bath should be advised where practicable. In
most cases, all that is necessary is to direct the patient to
confine himself to his room for a few days; to keep the head
and feet warm ; to swallow some hot drinks frequently during
the day ; to use linen pocket-handkerchiefs, not silken or cotton, and to change them frequently ; and to smear the upper
lip with salve to protect it from the acrid secretion. In the
latter stages, a long walk in the open air or even an occasional pinch of snuff hastens the cure. In infants at the
breast, the nostrils should be cleared by syringing them with
warm water, and they should be fed by the spoon or bottle
so long as the obstacle to sucking continues.

In the treatment of *chronic* nasal catarrh the cachexia,
whatever it may be, should be attended to. Local remedies
are also of the utmost importance. The most effective is penciling the swollen mucous membrane with the following solution :

149.   R.   Argenti nitratis,   gr.iv-xxx.
            Aquæ destillatæ,   f.℥j.   M.

Or, cauterizing with the lunar caustic in substance. The
following, to be used as a snuff, is in great repute :

150.   R.   Hydrargyri chloridi mitis,
            Hydrargyri oxidi rubri,   aa   gr.xij.
            Sacchari albi,   ℥ss.   M.
       For one powder.

When the discharge is offensive, and the above treatment fails, the following may be tried:

151. R. Iodinii,   gr.ij-iv.
          Potassii iodidi,   gr.iv-viij.
          Aquæ,   f.℥vj.   M.
To be used as an injection into the nostrils.

PROF. J. LEWIS SMITH, M. D., NEW YORK.

In children ordinary attacks of this affection require little treatment beyond keeping the bowels open, soaking the feet in mustard water, and having the body warmly clothed. Friction with camphorated oil over the nose is of some benefit. In attacks which commence with greater severity, an emetic of syrup of ipecacuanha given early will moderate the inflammation, and may prevent the occurrence of bronchitis. Afterwards a simple diaphoretic mixture should be administered, such as the following:

152. R. Syrupi ipecacuanhæ,   f.℥ij.
          Spiritûs ætheris nitrosi, f.℥j.
          Syrupi simplicis,   f.℥ij.   M.
One teaspoonful every three hours to a child of six months.

In place of sweet spirits of nitre *acetate of potash* may be employed in the dose of one to two grains for infants. If there is decided febrile reaction from half a minim to two minims, according to the age, of *tincture of digitalis* may be given in each dose.

In pseudo-membranous coryza the laryngitis which usually accompanies this affection demands the first attention. The frequent injection of a solution of chlorate of potash in water, several times a day, subdues the inflammation and removes the collection of mucus and pus. Or the following injection may be employed:

153. R. Pulveris aluminis,   gr. xvj-xx.
          Aquæ,   f.℥iv.   M.
The bromine solution (F. 176), diluted as directed, will also be found useful when injected into the nostrils.

Chronic coryza should be treated by tonics and by alteratives directed to the cachexia, which may be present. Together with such constitutional treatment (F. 153) may be injected into the nostrils, or a solution of nitrate of silver (gr. iij-v to f.ʒj.) An excellent formula for application to parts which can be reached by a camel's hair pencil is

154. R. Pulveris zinci oxidi, ʒj.
          Glycerinæ, f.ʒj. M.
    To be applied three or four times a day.

DR. L. WALDENBURG, OF BERLIN.

155. R. Ammoniæ muriatis, gr. iv.
          Aquæ destillatæ, f.ʒj. M.

For inhalation by atomizer, in that form of coryza known as dry snuffles. When the mouth is closed and the proper inclination given to the head, the nebulized spray can readily be inspired into the nostrils.*

Solutions of common salt may also be employed in the same manner (F. 160).

ACUTE LARYNGITIS.

WILLIAM AITKEN, M. D., EDIN.

If there be time in a case of acute laryngitis, commence by the inhalation of the steam of boiling water, as much by the nostrils as possible. Leeches, followed by hot fomentations, may be applied to the throat.

If benefit do not follow these remedial measures, tracheotomy ought not to be delayed. The air for respiration ought to be warm and moist, and plenty of it, through a large-sized canula, the orifice of which must be kept clear of secretion.†

*Cohen on Inhalation: p. 102.
†The Science and Practice of Medicine, Am. Ed., vol. ii. p. 832.

INHALATION.

J. M. DA COSTA, M. D.

156. R. Ammoniae muriatis,   gr. x-xx.
    Aquae destillatae,   f.℥j.   M.

Use with any form of steam atomizer throwing a fine spray. In *laryngeal catarrh*, acute as well as chronic. The dose best borne is not above ten grains to the ounce, although as much as two drachms to the ounce have been employed.*

LONDON HOSPITAL.

VAPOR BENZOINI.

157. R. Tincturae benzoini comp.,   f.℥j.
    Aquae bullientis,   f.℥x.   M.

Let the vapor be inhaled frequently.

VAPOR IODI.

158. R. Tincturae iodinii,   ℳ̃xl.
    Aquae bullientis,   f.℥x.   M.

To be inhaled frequently.†

THOMAS HAWKES TANNER, M. D., F. L. S., ETC., LOND.

159. R. Acidi hydrocyanici diluti,   ℳ̃xv.
    Spiritûs chloroformi,   f.℥iij.
    Aquae bullientis,   f.℥viij.   M.

The patient should frequently inhale the steam from medicated boiling water, and in the interval it will prove advantageous for him to wear a respirator. He is to be closely watched, kept very quiet, and not allowed to talk. The air of the room must be made warm and moist.‡

DR. L. WALDENBURG, OF BERLIN.

160. R. Sodii chloridi,   gr. iv-x.
    Aquae destillatae,   f.℥j.   M.

For inhalation in acute laryngeal catarrh by means of a nebulizer.§

*Inhalation in the Treatment of Diseases of the Respiratory Passages, p. 52.
†Squire's Pharmacopœia of the London Hospital, 2d Ed., p. 184.
‡The Practice of Medicine, p. 335.
§ Inhalation: Its Therapeutics and Practice. By J. Solis Cohen, M. D., (L. & B.) 1869, p. 111.

CROUP.

HERMAN BEIGEL, M. D., ETC., LONDON.

Our author recommends in a case of croup the use of the following medicated sprays, given in the order of their value. (They are applied by means of the atomizer.)

161. ℞. Aquæ calcis,   f.ℨj.

For one inhalation, lasting about a quarter of an hour, and to be repeated every two hours as long as bad symptoms are present.

(The method of Dr. A. Geiger, of Dayton, Ohio, is to pour hot water on unslacked lime in a pitcher, and to have the patient inhale the vapor as it rises*).

162. ℞. Acidi tannici,   gr. ij-xx.
      Aquæ,   f.ℨj.   M.

For one inhalation, to last fifteen to twenty minutes.

163. ℞. Potassii bromidi,   gr. v-x.
      Aquæ,   f.ℨj.   M.

This inhalation at an early stage of the disease will often be found to arrest the symptoms.

With these inhalations the administration of emetics or other remedies may and must be combined if considered necessary.†

MEREDITH CLYMER, M. D., ETC., NEW YORK CITY.

Our author believes that blood-letting, either general or local, is generally harmful in the treatment of croup. The safest and best emetic is sulphate of copper, in half or one grain doses, every fifteen minutes, till vomiting occurs. Afterwards give the following:

164. ℞. Potassæ chloratis,   Ɖij.
      Potassii iodidi,   gr. viij.
      Tincturæ opii camphoratæ,   ♏xl.
      Liquoris potassæ,   ♏xij.
      Aquæ,   f.ℨij.   M.

A dessertspoonful in water every second or third hour.

---

*Philadelphia *Medical and Surgical Reporter*, March 10th, 1866, p. 195.
†Half-yearly *Compendium of Medical Science*, Jan., 1869, p. 93.

Brush the throat and upper part of the larynx with the
following :

165. R.   Tincturæ ferri chloridi,   f.ʒj.
          Aquæ,   f.ʒij.   .M

Beef tea, wine, quinine, and iron are generally required.

INHALATION.

J. M. DA COSTA, M. D.

166. R.   Extracti hyoscyami fluidi,   ♏iij-x.
          Aquæ destillatæ,   f ʒj.   M.

DR. FELIX VON NIEMEYER, PROFESSOR OF PATHOLOGY AND THERAPEUTICS, ETC., UNIVERSITY OF TUBINGEN.

167. R.   Cupri sulphatis,   gr. x-xv.
          Aquæ,   f.ʒij.   M.

A large teaspoonful to be taken every five minutes until
vomiting sets in.

In regard to the employment of emetics in croup our author thinks that the revulsive action through which they are
supposed to exert an influence upon the disease is altogether
problematic. Still less can help be looked for from their diaphoretic effect. *They are only indicated where obstructing
croup-membranes play a part in producing the dyspnœa, and
when the child's efforts at coughing are insufficient to expel them.
Impeded expiration is* an indication for their employment.

Sulphate of copper is preferable to tartar-emetic or ipecacuanha. It should be given in full doses ; in small ones it is
uncertain, and more apt to operate as a poison. The more
complete the remission after the vomiting, the more of the membrane thrown out, so much the more reason for repeating the
emetic, should the *impeded expiration* recur. If there should
be no remission, should no croup-membrane be expelled, or if
the expiratory act be free from impediment, the repetition of
the emetic is contra-indicated.

The *application of cold* deserves a full trial, in the shape of cold compresses quickly changed, laid upon the throat of the child as soon as the signs of croupous laryngitis appear. In families where they are not afraid to use this treatment, much happier results will be obtained than in houses in which the prejudice against it is not to be overcome.

If the bowels be confined administer a clyster so that the diaphragm may have room to act. The best is a cold one, as follows:

168. ℞. Acidi acetici diluti, f.ℨij.
Aquæ, f.ℨvj. M.

If, however, there is no remission, notwithstanding the employment of the emetic and the cold application, apply the following concentrated solution, at intervals of several hours, to the entrance of the glottis:

169. ℞. Argenti nitratis, ℨss.
Aquæ destillatæ, f.ℨij. M.

Dip a curved-rod of whalebone, with a small sponge made fast to its lower end, into this solution, press down the tongue of the child, and endeavor to reach the entrance of the glottis with the sponge. There the sponge is immediately compressed by the muscular contraction which takes place, whereby certainly a portion of the liquid, if only a small one, arrives at the larynx. Administer also half a grain of calomel every two hours.

Should this treatment remain without effect, proceed at once to tracheotomy.

Besides treating the dyspnœa upon the principles given above, it may be necessary to relieve the paralytic symptoms due to blood-poisoning by carbonic acid. For this purpose the powerful stimulus obtained by pouring cold water upon the child while in a warm-bath is of great service. Lose no time in making use of it, the moment the child begins to grow drowsy, the skin to cool, the sensorium to be benumbed, or as soon as the emetics fail to act. A few gallons of cold water, poured from a moderate height, over the head, nape and back of the child, almost always cause it to revive for a while and to cough vigorously. Thus, sometimes after the bath, masses of exudation are expelled. Other stimulants, such as camphor or musk, are much less effective, and ought not to be employed save when insuperable objections are opposed to the cold effusion. They should be given in large doses, immediately prior to the emetic. The following formula may be used:

170. R. Camphorae,   gr. x.
          Ætheris acetici,   f.ʒiij.   M.
     Ten to fifteen drops to be given every quarter of an hour.

171. R. Moschi,   gr. iv.
          Sacchari albi,   ʒj.   M.
     Divide into vj powders. Direct one every hour, or half
     hour.*

DR. JOHANN SCHNITZLER.

INHALATION.

172. R. Potassii bromidi,   gr. x.
          Aquæ destillatæ,   f.ʒj.   M.
     To be inhaled by means of a nebulizer in pseudo-membranous croup.†

PROF. J. LEWIS SMITH, M. D., ETC., NEW YORK.

173. R. Potassæ chloratis,   ʒj.
          Ammoniæ muriatis,   ϶ij.
          Syrupi simplicis,   f.ʒj.
          Aquæ,   f.ʒij.   M.
     One teaspoonful every twenty minutes to a half hour, or in cases not severe every two hours. This should be continued regularly night and day until the cough becomes looser, or until it is evident, if the case is unfavorable, that it can be of no service.

The atmosphere the child breathes should be constantly loaded with moisture, without, however, that degree of heat

---

*Text-Book of Practical Medicine. Am. Ed., 1869, vol. I, p. 30.
†Cohen on Inhalation, p. 115

which would add materially to the discomfort of the patient or attendants. The temperature should be of 75° or 80°.

Besides the nitrate of silver, three other substances have been used of late years for the topical treatment of the throat, which appear to be more effectual in removing the pseudo-membrane, and controlling the inflammation. One is liquor ferri subsulphatis; the other, carbolic acid, and the third, bromine. The following formulæ may be used:

174. R. Liquoris ferri subsulphatis, f.ʒj
    Glycerine, f.ʒss. M.
175. R. Acidi carbolici fluidi, f.ʒj.
    Aquæ, f.ʒvj. M
176. R. Brominii, ʒij.
    Potassii bromidi, gr. xlv.
    Aquæ, f.ʒj. M.

This is called the bromine solution; but it must be considerably diluted for use. Twenty-four to forty drops should be added to an ounce of water for application to the fauces or larynx. Our author most highly recommends the subsulphate of iron solution.

APHONIA.

J. M. DA COSTA, M. D., PHILADELPHIA.

177. R. Ammoniæ muriatis, ʒss.
    Syrupi pruni virginiani, f.ʒiij. M.

A teaspoonful ter die in intermittent aphonia, together with

178. R. Strychniæ sulphatis, gr. 1-50
In granule, ter die.

Used in a case of intermittent aphonia, in which the voice was lost for an hour at a time several times a week. A catarrhal condition in this woman, a school teacher, lead to a weakening of the cords so that any over-exertion brought about temporary inability to generate distinct voice.

The local application of sulphate of zinc to the cords is productive of permanent good in such cases. The following formulæ may be used for

INHALATION.

179. R. Zinci sulphatis,   gr. j-vj.
          Aquæ destillatæ,   f.ℨj.   M.

180. R. Ferri sesquichloridi,   gr. ⅓-ij.
          Aquæ destillatæ,   f.ℨj.   M.

As a weak inhalation in hysterical aphonia.

DR. FIEBER.

181. R. Tincturæ opii,   gtt. vj.
          Zinci sulphatis,   gr. vj.
          Aquæ destillatæ,   f.ℨj.   M.

In laryngo-tracheal catarrh, our author reports a case with almost complete aphonia, in which the voice soon returned under the inhalation of this solution.*

In aphonia from paralysis of the vocal cords local Faradization should be employed.

M. O. REVEIL, OF PARIS.

182. R. Pulveris benzoinii,   ℨss.

Place a portion on some live coals, and inhale the vapor, by deep inspirations in cases of aphonia and hoarseness.†

PROF. J. LEWIS SMITH, M. D., NEW YORK.

Chronic laryngitis, dependent on syphilis or tuberculosis, requires specific treatment; local measures have but little effect. The chronic laryngitis, occurring in children in good general health, sometimes resulting from an acute attack, is an obstinate affection. The patient should be warmly clad, and every effort made to guard against taking cold. The most satisfactory treatment is the application of tincture of iodine upon the neck, directly over the larynx, and in some

---

*Cohen on Inhalation; p. 142.
†Formulaire Raisonné des Médicaments Nouveaux, etc., Deuxieme édition, p. 249.

cases of a solution of nitrate of silver, ten or twenty grains to the ounce, to the fauces, so that if possible some of it may enter the larynx. Little benefit is derived from stimulating expectorants. The following recipe has proved beneficial in a number of cases.

183. ℞. Extracti cobebæ fluidi, ℳ xl.-f.ʒj.
Syrupi simplicis, f.ʒijss. ℳ.
A teaspoonful three or four times daily.

INHALATION.

DR. L. WALDENBURG, OF BERLIN.

184. ℞. Sodii chloridi, gr. ij.
Aquæ destillatæ, f. ʒj. ℳ.

Used with benefit in a case of aphonia in a patient affected with laryngitis and tuberculosis.*

ASTHMA.

J. M. DA COSTA, M. D., PHILADELPHIA.

185. ℞. Spiritûs ætheris compositi,
Extracti valerianæ fluidi, aa f.ʒj.
Tincturæ lobeliæ, f.ʒss.
Potassæ chloratis, ʒjss.
Syrupi tolutanus, f.ʒj. ℳ.
A dessertspoonful in water three times a day.

Direct, also, the following stimulating liniment:

186. ℞. Chloroformi, f.ʒss.
Olei terebinthinæ, f.ʒj.
Spiritûs rosmarini, f.ʒjss. ℳ.
To be rubbed on the chest several times a day.

187. ℞. Potassii iodidi, ʒij.
Morphiæ sulphatis, gr. ¾.
Tincturæ scillæ,
Tincturæ lobeliæ,
Syrupi, aa f.ʒj. ℳ.

*Cohen on Inhalation; p. 142.

A teaspoonful ter die in asthma, with emphysema and chronic bronchitis. ℞

188. ℞. Zinci valerianatis, Əij.
Extracti belladonnæ, gr. j. M.
For twenty pills. One ter die in *nervous* asthma.

In the treatment of the paroxysms of asthma all nauseants have a certain influence in relaxing the bronchial spasms. Lobelia is particularly serviceable, because it acts both as a nauseant and expectorant.

189. ℞. Tincturæ lobeliæ,
Tincturæ hyoscyami,
Spiritûs ætheris compositi,
Syrupi tolutanus, aa f.ʒj. M.
A teaspoonful in water every ¦half hour during the paroxysm, until some effect is produced in the breathing, and then every hour or two.

Strong coffee is also sometimes of service in averting a paroxysm, (Dr. NIEMEYER directs two ounces of Mocha to the cup.)

The fumes of nitre paper often give relief.

190. Take some ordinary blotting paper, dip it into a concentrated solution of nitrate of potassa and allow it to dry. When wanted for use, ignite it in an open vessel, covered with a newspaper made into a cone, so that the fumes may be inhaled.

In order to prevent the recurrence of the paroxysms too great care cannot be taken to inquire into the state of all the functions. In perhaps one-half of the cases asthma is not a disease of the lungs, but a reflected trouble.

In cases of stomach origin,

191. ℞. Pilulæ hydrargyri, gr. ij.
At night; to be followed by half an ounce of Rochelle salts in the morning. Afterwards direct arsenic combined with gentian or belladonna; as,

192. ℞. Liquoris potassæ arsenitis, f.ʒj.
Tincturæ gentianæ compositæ, f.ʒiij. M.
A desertspoonful, ter die.

5

INHALATION.

M. FAURE.

193. R. Aquæ ammoniæ,   f.℥ss.

To be placed in a bowl, and the vapor inhaled, taking care however, to close the nostrils. The inhalation is to be continued for fifteen minutes, and to be repeated four times a day.*

C. HANDFIELD JONES, M. B. CANTAB., F. R. C. P., LOND.

The treatment of asthma is eminently that of a neurosis. It consists first in the removal of all causes of irritation, such as catarrhal inflammation of the bronchi, dyspeptic disorder, unwholesome diet and unsuitable climate, etc.; and secondly, in the use of various appropriate tonics and sedatives. Arsenic is sometimes of much advantage.

194. R. Liquoris potassæ arsenitis,   ℥iv-v.

For one dose ter die, either alone or combined with an ordinary cough mixture.

195. R. Extracti cannabis indicæ,   gr. v.

For ten pills.

This drug occasionally proves very useful, one pill immediately checking the spasm.

HYPODERMIC INJECTION.

196. R. Liquoris opii sed.,   ℥x.

For one subcutaneous injection into the left front of the chest.

197. R. Atropiæ sulphatis,   gr. 1-35.
         Aquæ,   ℥x.   M.

For one hypodermic injection in the vicinity of the left vagus nerve.†

DR. FELIX VON NIEMEYER, PROFESSOR UNIVERSITY OF TUBINGEN.

Among medicaments especially in repute for the prevention

---

* Handfield Jones on Functional Nervous Disorders. Am. Ed., p. 244
† Functional Nervous Disorders. Am. Ed., Phila., 1867. p. 243.

of new paroxysms, and for the radical cure of asthma, quinine
stands first. The shorter and more regular the intervals of
the attack, so much the more is to be expected from this drug.
It is unsuitable when the pause between the seizures are very
long or irregular in their occurrence. In such cases we must
have recourse to other remedies from the list of the so-called
nervines.

As a rule, the metallic nervines are to be preferred before
tincture of valerian, assafœtida, castor or camphor. The following may be used:

198. R. Ferri carbonatis, ℥j.
Syrupi, q. s. M

For xx pills. One ter die.

199. R. Zinci oxidi, gr.xxv
Syrupi, q. s. M.

For xx pills. One ter die.

200. R. Argenti nitratis, gr. iij-iv.
Confectionis rosæ, q. s. M.

For xxiv pills. One ter die.

DR. J. S. MONELL, OF NEW YORK.

Our author recommends *forced expiration and inspiration
for the relief of spasmodic* asthma.

He directs that all the air be expired that it is possible for
the patient to do, and not to inspire until it is found absolutely necessary. Then carry inspiration to its fullest capacity,
and retain with great effort for many seconds. This act of
forced expiration, waiting, thorough inspiration, and again
waiting, should be continued for some fifteen minutes, when
it will be found that the spasm is relieved. It requires great
exertion on the part of the patient to perform this act. The
first attempt at retaining the inspired air during the asthmatic
attack will cause the patient to think he cannot continue it,

but perseverance will soon delight him with relief from the spasm.*

DR. PRIDHAM, OF BIDDEFORD, IN DEVONSHIRE, ENG.

Our author has been very successful in the treatment of asthmatic cases by means of a duly regulated diet and sedatives during the intervals of the paroxysms.†

The secretions from the bowels are, first of all, to be corrected by the following pill at bed-time, succeeded by a saline aperient in the morning:

> 201. R. Pilulæ aloes cum myrrhâ,   gr. iij.
>   Pilulæ hydrargyri,   gr. j.
>   Extracti taraxaci.   gr. ij.
>   Extracti sthrmonii,   gr. ss.   M.
>   For two pills.

Or, giving every alternate night, in the form of a pill:

> 202. R. Pilulæ hydrargyri,   gr. iv
>   Pulveris ipecacuanhæ,   gr. j.   M.
>   For one pill.

And on the following morning:

> 203. R. Misturæ sennæ compositæ,   f.ʒj.
>   Magnesiæ bicarbonatis,   gr. x.
>   Sodæ bicarbonatis,   gr. viij.   M.
>   For one draught.

During the day small doses of *compound rhubarb powder*, of which the following is the formula: (Br. Phar.)

> 204   R. Pulveris rhei radicis.   ʒij.
>   Magnesiæ,   ʒvj.
>   Pulveris zingiberis,   ʒj.
>   To be mixed thoroughly and passed through a fine seive.
>   Dose—20–60 grains.

After having thus attended to the general secretions for about ten days, the strict dietary system is to be commenced.

He restricts his patients at first to two ounces of fresh meat, with as much dry bread for dinner at one P. M., and the same for supper at seven ; allows a cup of tea with cream, and dry bread, in the morning; and for drink, weak brandy or whisky and water, which is not to be taken till three hours after animal food. Rest is also enjoined for the same period, though air and exercise are recommended. The following sedatives are also to be given :

205. R. Extracti conii, 3j.
Extracti cannabis indicæ, gr. v. M.

For twenty pills. One to be taken four times a day, at the hours of seven, twelve, five, and ten. Gradually the dose of extract of conium is to be increased to gr. v., and that of Indian hemp to gr. j., five times a day.

DR. HYDE SALTER, LONDON.

206. R. Potassæ nitratis. 3iv.
Aquæ, Oss. M.

This is the solution which Dr. S. directs to be used in the making of nitre paper. Red blotting paper should be used.

The following formula for the preparation of a paper to be burnt for the relief of asthma is given in the *Journal de Pharmacie et de Chimie :*

Take four ounces of white paper, and allow it to macerate in warm water until reduced to a uniform paste. Then press out the greater portion of water and mix the residue in a mortar with the following powder: Nitrate of potash, 3ij; myrrh and olibanum aa 3ijss; belladonna, stramonium, digitalis, aa gr. x. When a uniform mass has been formed, roll out into sheets a line or so thick ; dry and cut into strips. This paper is said to burn less quickly than the ordinary nitre paper, and to be more effective.*

---

* *Half-Yearly Compendium of Medical Science*, January, 1869, p. 71.

THOMAS HAWKES TANNER, M. D., F. L. S., LONDON.

207. R. Potassii iodidi,   ℨijss-℈iv.
         Spiritûs ammoniæ aromatici, f.ʒj.
         Tincturæ belladonnæ,   ℳl-f.ʒijss.
         Tinct. cinchoniæ comp.,   f.ʒij.
         Aquæ menthæ piperitæ,   f.ʒj.   M.
A dessertspoonful, in water, three times a day.

In some cases of asthma the author has found remarkable
benefit from this formula.   It requires to be persevered with
for some weeks; the patient being watched lest it impoverish
the blood and produce purpura or boils, or even a carbuncle.
If there be constipation order

208. R. Pilulæ rhei compositæ,
         Extracti conii,   aa   gr. v.
For two pills.   To be taken at bed time.

DR. WISTINGHAUSEN.

INHALATION.

209. R. Liquoris potassæ arsenitis, gtt. x-xv-xx.
         Aquæ destillatæ,   f.ʒj.   M.
For inhalation, by means of atomizer, once or twice a day†.

HÆMOPTYSIS.

WILLIAM AITKEN, M. D., EDIN.

210. R. Potassæ bitartratis,   ʒj.
         Pulveris opii,   gr. ij-iv.   M.
For eight powders.   One every four or six hours.

211. R. Plumbi acetatis,   ℈j-ʒj.
         Pulveris opii,   gr. x.   M.
For xx pills.   One every four or six hours.

When hæmoptysis is connected with amenorrhœa preparations of iron often succeed when the above remedies fail.
Thus :

---

†Cohen on Inhalation, p. 155.

212. ℞. Ferri sulphatis,   ℈ij.
          Magnesiæ sulphatis,   ℥ijss.   M.
For xx powders, one ter die.

This will often restore the menstrual secretion and cure the
hæmoptysis.   Indeed, it is in this form of amenorrhœa that
iron is most successful.

213. ℞. Spiritûs terebinthinæ,   f.℥ss.
For one dose, if the bleeding is unattended with vascular
excitement.

J. M. DA COSTA, M. D., PHILADELPHIA.

214. ℞. Acidi gallici,   gr. xx.
For one powder. In acute hemoptysis. To be repeated
every ten minutes until hemorrhage ceases.

215. ℞. Cupri sulphatis,   gr. ½.
          Ferri sulphatis,   gr. ij.
          Extracti hyoscyami,   gr. j.   M.
For one pill, ter die.

In persistent slight pulmonary hemorrhage. Sulphate of
copper in such cases seems to control the circulation within
the lungs, and to arrest a tendency to bleeding. This action
is not due to its nauseating effect, for, in this dose, it does not
nauseate. The administration of cod-liver oil should be post-
poned until the hæmoptysis is gotten rid of.

INHALATION.

216. ℞. Ferri sesquichloridi,   gr. ij-x.
          Aquæ destillatæ,   f.℥j.   M.
For inhalation by atomizer.

217. ℞. Liquoris ferri subsulphatis,   ♏x-xl.
          Aquæ destillatæ,   f.℥j.   M.
For inhalation by atomizer.

218. ℞. Acidi tannici,   gr. x-xx.
          Aquæ destillatæ,   f.℥j.
For inhalation by atomizer.

219. ℞. Pulveris aluminis,   gr. xxx.
          Aquæ destillatæ,   f.℥j.   M.
For inhalation by atomizer.*

---

*Da Costa on Inhalation: p. 51

DR. HORACE DOBELL, SENIOR PHYSICIAN TO THE ROYAL HOSPITAL FOR DISEASES OF THE CHEST, LONDON.

220.　R.　Extracti ergotæ fluidi,　　　f.ʒij.
　　　　　Tincturæ digitalis,　　　　　f.ʒij.
　　　　　Acidi gallici,　　　　　　　ʒj.
　　　　　Magnesiæ sulphatis,　　　　ʒv.
　　　　　Acidi sulphurici diluti,　　　f.ʒj.
　　　　　Infusi rosæ compositi,　　　f.ʒvj.　　M.

Sig.—Two tablespoonfuls every three hours until the hemorrhage is arrested.

In spite of the fashionable outcry against complicated prescriptions, Dr. Dobell recommends the above as the most efficacious and the most rational combination of remedies for a case of profuse tubercular pulmonary hemorrhage. In any given case, either of the ingredients may be omitted, if the symptoms indicate that it is not required, or that it has already done its duty. The object of the ergot is to contract the vessels; of the digitalis, to steady the heart; of the gallic acid, to clot the blood; of the epsom salts, to relieve the congestion; and of the dilute sulphuric acid, to assist the rest.

DR. FELIX VON NIEMEYER, PROF. UNIVERSITY OF TUBINGEN.

221.　R.　Copaibæ,
　　　　　Syrupi,
　　　　　Aquæ menthæ piperitæ,
　　　　　Spiritûs vini rectificatæ, aa f̄ʒj.
　　　　　Spiritûs ætheris nitrosi,　　f.ʒss.　　　M.

A dessertspoonful every two to four hours.

A formula much in use in very obstinate hæmoptysis.*

ACUTE BRONCHITIS.

WILLIAM AITKEN, M. D., EDIN.

When the symptoms of a "common cold" first express themselves, and even when the trouble has extended to the

*Text Book of Practical Medicine; Am. Ed. Vol. 2; p. 346.

chest, as indicated by the hoarseness and tendency to cough, the disease may at once be subdued in a healthy person by a full stimulant, but not narcotic dose, of opium or morphia, *i. e.*,

222. R. Pulveris opii,   gr. j.
                     vel
223. R. Liquoris morphiæ sulphatis, f.ʒij.
      For one dose at bed-time.

Or by v grains of ammoniæ carbonatis; or by x to xx grains of ammoniæ muriatis; or by an alcoholic diaphoretic drink; or, if the appetite is unimpaired, by a full supper, followed by a moderate amount of some alcoholic stimulant.

If such remedies are delayed too long the object to be aimed at most is to induce a copious perspiration, and a continued action of the skin and kidneys. From the frequent inhalation of steam great benefit is derived.

If the disease shows a disposition to pass into the chronic stage, the following may be administered, and will generally facilitate expectoration and relieve the dyspnœa, viz:

224. R. Ammoniæ carbonatis,   gr. v.
          Tincturæ benzoini comp.,   f.ʒss.
          Tinct. cinchoniæ compositæ. f.ʒjss.   M.
      For one dose.*

DR. JAMES COPLAND.

225. R. Liq. ammoniæ acetatis,   f.ʒj.
          Spt. ætheris nitrosi,   f.ʒiij
          Vini antimonii,   f.ʒijss.
          Misturæ amygdalæ,   ad f.ʒviij.   M.

One or two tablespoonfuls every third or fourth hour in the treatment of primary or simple bronchitis. The bowels should be moderately opened by a small dose of calomel or blue pill, with antimonial powder at night, and a gentle aperient in the morning. If the patient be aged, delicate, or vitally de-

---

†Science and Practice of Medicine. Am. Ed., 1858, vol. ii, p. 701.

pressed, the antimony may be omitted from the prescription, and a proportion of the infusion of cinchona, or of the decoction of senega added. The dose of the antimony may, of course, be increased if indicated. When the fever is considerable, or the patient complains of soreness or pain in the chest, a mustard poultice may be applied over the sternum, or the following terebinthinate embrocation employed:

226. R. Liuimenti saponis.
"   terebinthinæ, aa f.ʒjss.
Olei olivæ,   f.ʒvij
Olei cajuputi,   f.ʒj.   M.

This embrocation, having been well shaken, should be sprinkled on two or three folds of flannel, or on spongeo-piline, and placed either over the thorax or between the shoulders.

J. M. DA COSTA, M. D., PHILADELPHIA.

227. R. Vini ipecacuanhæ,   f.ʒij.
Liquoris potassæ citratis,   f.ʒiv.
Tincturæ opii camphoratæ,
Syrupi acaciæ,   aa f.ʒj.   M.
A tablespoonful ter die in the first stage of ordinary acute bronchitis

228. R. Morphiæ acetatis,   gr j.
Potassæ acetatis,   ʒiij.
Liquoris ammoniæ acetatis,   f.ʒiij.
Syrupi tolutanus,   f.ʒj.   M.
A dessertspoonful every third hour. A useful diaphoretic alkaline mixture.

229. R. Ammoniæ carbonatis,   gr. xvj.
Spiritûs ætheris compositi,   f.ʒjss.
Syrupi tolutanus,
Aquæ,   aa f.ʒj.   M.
A teaspoonful every two hours; a stimulating expectorant for a child a year old, affected with bronchitis of two weeks' standing.

Counter-irritation to be applied to the chest by means of weak mustard plasters (one part of mustard to four of Indian meal). Also, if the child be much debilitated, 15 drops of brandy every four hours. When the child is seen frequently, so that the effect may be watched, there is no better treatment than relieving the lung mechanically by emetics. Hoffman's anodyne, in the above recipe, acts as a diaphoretic and quieting agent, which latter influence would not be obtained from sweet spirits of nitre.

230. R. Syrupi ipecacuanhæ, f.ʒss.
Liquoris potassæ citratis, f.ʒijss.
Misturæ glycyrrhizæ comp., f.ʒj. M.

A teaspoonful every three hours; for a child two years of age. Afterward, when the disease passes into second stage, to be changed to

231. R. Syrupi scillæ, f.ʒij.
Tincturæ opii camphoratæ, f.ʒij. M.

Thirty drops four times a day.

232. R. Syrupi ipecacuanhæ, f.ʒj.

A teaspoonful every ten minutes until vomiting is produced; to be repeated every second day. For ordinary acute bronchitis in a child a year old. Together with

233. R. Ammoniæ carbonatis, ʒss.
Syrupi senegæ, f.ʒss.
Syrupi tolutanus, f.ʒj.
Aquæ, f.ʒijss. M.

A teaspoonful, ter die.

PROF. A. P. DUTCHER, M. D., CLEVELAND, OHIO.

TREATMENT OF ACUTE BRONCHITIS IN CHILDREN.

Emetics and mercurials are useful in nearly every stage of the malady. Among the first-class, ipecac. ranks the highest, given in the form of powder or syrup. When the fever is high the patient should be placed in a warm bath, after which sufficient ipecac. may be given to produce vomiting, particularly if there is much mucus in the bronchial tubes. In most cases, however, it will be found more useful in smaller doses, combined with calomel and soda, viz.:

234. ℞. Hydrargyri chloridi mitis,   gr. vj.
         Pulveris ipecacuanhæ,   gr. iij.
         Sodæ bicarbonatis,   gr. xxiv.   M.
    For six powders.

One to be given every three hours (to a child two years old) until the bowels are freely moved. After which they may be given at longer intervals, and the calomel may be reduced to one-fourth the quantity, or omitted altogether.

Sometimes after the bowels have been freely moved our author substitutes the following with the most happy effect :

235. ℞. Syrupi ipacacuanhæ,   f. ℥ss.
         Tincturæ digitalis,   f. ℥ss.
         Spiritûs ætheris nitrosi, f. ℥jss.
         Tincturæ opii camphoratæ, f. ℥ij.
         Syrupi simplicis,   f. ℥j.   M.
    A teaspoonful every four hours.

In cases which do not yield at once to the remedies just named, resort should be had to counter-irritants to the chest. Our author considers the following liniment preferable, in very young children, to either the mush jacket or mustard poultices :

236. ℞. Pulveris camphoræ,   ℥ss.
         Olei terebinthinæ,   f. ℥ij.
         Olei olivæ,   f. ℥iij.   M.

This is to be applied to the breast by means of flannel thoroughly saturated with it, which should be covered with a piece of oil silk and securely kept in place by a thin roller. In older children where the disease has extended to the smaller bronchia, and appears to be obstinate, a blister should be applied.

In the management of children, affected with acute bronchitis, there is one rule which should never be neglected. *Never allow the patient to remain too long in one posture.* The child should not be suffered to lie in one position more than two hours.

ACUTE BRONCHITIS IN ADULTS.

If the patient be robust, and the attacks very acute, attended with a frequent and hard pulse, hot and dry skin, costive bowels, laborious breathing, and a troublesome, dry cough, our author recommends the taking of twelve or fifteen ounces of blood from the arm. If bleeding has no influence in curing the disease, the wonderful power it has in mitigating the patient's sufferings, places it far beyond every other known therapeutical agent.

After the bleeding, if the skin remains hot and dry, the patient may be treated to the alcoholic fumigating bath and one of the following powders, given every four hours until the bowels are freely moved:

237. R. Podophyllin,   gr. j.
          Hydrargyri chloridi mitis,   ℈j.
          Potassæ nitratis,   ℥ss.
          Pulveris ipecacuanhæ,   gr. viij.   M.
      For four powders.

If, after the bowels have been freely moved, the skin should remain hot and dry, the patient may again be subjected to the fumigating bath, and a teaspoonful of the following mixture administered every two hours, until free perspiration and expectoration are produced:

238. R. Tincturæ veratri viridi,   f.℥ss.
          Syrupi scillæ compositi,   f.℥iij.
          Spiritûs ætheris nitrosi,   f.℥ss.
          Extracti lobeliæ fluidi,   f.℥ij.
          Tincturæ opii camphoratæ,   f.℥ss.   M.

If this should produce nausea or vomiting after a few doses, it will commonly prove very beneficial. After free expectoration has been produced, it may be given at longer intervals, or superseded by the following:

239. R. Tincturæ sanguinariæ,   f.ʒij.
       Vini ipecacuanhæ,   f.ʒj.
       Morphiæ sulphatis,   gr. ij.
       Syrupi simplicis,   f.ʒj.   M.
A teaspoonful every six hours.

To relieve cough and produce rest at night, give the patient eight or ten grains of Dover's powder early in the evening.

If the disease still proves obstinate, and if the dyspnœa and cough still continue, the patient complaining of great weight upon the chest, with a deep, burning pain just under the sternum, apply a blister immediately over the seat of the difficulty. It will almost invariably relieve all the symptoms, cut the disease short, and secure a safe and speedy convalescence.

After the disease has passed its climax, very little medical treatment is demanded. If the cough should remain troublesome and the expectoration scanty and tenacious, our author is in the habit of prescribing the following:

240. R. Ammonii bromidi,
       Ammonii iodidi,   aa ʒij.
       Morphiæ sulphatis,   gr. ij.
       Syrupi phellandrii aquatici
         comp.,   f.ʒviij.   M.
A dessertspoonful every six hours.

TREATMENT OF ACUTE BRONCHITIS IN THE AGED.

Vegetable emetics in small doses and expectorants, especially those of a stimulating nature, are the most valuable remedies in this case. Senega is the best of the vegetable expectorants, and the decoction the best form.

241. R. Decocti senega (U.S.P.),   f.ʒviij
Two fluid ounces to be administered every four or six hours.

If the patient be weak, add ammoniæ carbonatis, gr. v, and quiniæ sulphatis, gr. ij, to each dose. A small glass of ale three or four times a day is a mild stimulant, expectorant and anodyne, much to be preferred to wine and ardent spirits. In bad cases of acute bronchitis our author never hesitates to blister; in mild cases he uses **F. 236.**

After the violence of the disease has passed, tonics and expectorants should be freely used. The following is a useful expectorant:

242. R. Syrupi scillæ,
 Syrupi senegæ,
 Syrupi ipecacuanhæ, aa. f.ʒss.
 Morphiæ sulphatis, gr. ij.
 Syrupi simplicis, f.ʒjss. M.
A teaspoonful every six hours.*

DR. LATHAM.

243. R. Confectionis rosæ,
 Mellis, aa. ʒjss.
 Tragacanthæ, gr. xxiv.
 Pulveris ipecacuanhæ, gr. vj.
 Syrupi tolutanus, f.ʒiij. M
A teaspoonful three or four times a day.†

PROF. J. LEWIS SMITH, NEW YORK.

244. R. Spts. ætheris nitrosi, f.ʒj
 Syrupi ipecacuanhæ,
 Olei ricini, aa. f.ʒij.
 Syrupi tolutanus, f.ʒvij. M.
One teaspoonful, in primary bronchitis, for an infant one year old, every two or four hours.

Another eligible formula is the following:

245. R. Syrupi ipecacuanhæ, f.ʒij.
 Potassæ acetatis, gr. xvj-ʒss.
 Aquæ anisi, f.ʒxiv. M.
Dose—one teaspoonful for an infant of six months. If there is decided febrile action, tincture of digitalis, one or two drops, according to the age, may be added to each teaspoonful.

---

*Philadelphia *Medical and Surgical Reporter*, Aug. 17, 1867, pp. 138, 139.
†Formulaire Raisonné des Médicaments Nouveaux, par O. Reveil, p. 251

In the majority of cases of infantile bronchitis this mode of treatment is preferred by Dr. Smith for the first few days, after which, if further medication is required, more sustaining or even stimulating medicines are proper.

For children over the age of three years, if the previous health has been good, and the bronchitis is primary, aconite or veratrum viride is often useful in the first stage of the inflammation. The following is a recipe for a child of five years:

> 246. ℞.   Tr. veratri viridi,   gtt. xij.
>   Syrupi scillæ comp.,   f.ʒij.
>   Syrupi tolutanus,   f.ʒxiv.   M.
> One teaspoonful every two or four hours; the medicine to be omitted or given at longer intervals if the frequency of the pulse is reduced.

The effect of cardiac sedatives should be carefully watched. In general they should be administered only during the first three or five days. But if the child is robust, with full and strong pulse, they may be continued longer.

As the active inflammation begins to abate, simple expectorant mixtures may be given, as syrup of squills or ipecacuanha, in spiritûs mindereri. At this stage of bronchitis in children it is often best to commence the use of stimulating expectorants, and they are required in nearly all cases of advanced bronchitis. A favorite prescription with Dr. Smith is the following:

> 247. ℞.   Ammoniæ carbonatis,   gr. xvj-xxiv.
>   Tincturæ sanguinariæ,   gtt. xxiv.
>   Syrupi senegæ,   f.ʒij.
>   Extracti glycyrrhizæ,   ʒss.
>   Aquæ,   f.ʒxiv.   M.
> Dose—One teaspoonful every three or four hours to a child of two years. If there is much restlessness, Dover's powder, paregoric, or syrup of poppies should be given with this mixture, or separately.*

* Diseases of Infancy and Childhood, 1869, p. 215.

ACUTE BRONCHITIS. 81

THOMAS HAWKES TANNER, M. D., F. L. S.

The patient should be confined to bed in a room of the temperature of from 65° (F.) to 70°, with the air kept moist. Beef-tea, milk, arrow-root or gruel, tea with milk and a mucilaginous drink ought to be allowed, such as

248. R. Misturæ acaciæ,
          Misturæ amygdalæ,   aa. Oss.
      To be mixed with half a pint of pure milk and sweetened with sugar-candy or honey. Then add one large tablespoonful of any liquor. Allow the whole to be taken during the day. Or,

249. Boil a large pinch of isinglass with a tumblerful of milk, half a dozen bruised almonds, and two or three lumps of sugar. To be taken warm once or twice in the day.

If there be indications of debility, white-wine whey, made according to the following formula, will prove a good restorative :

250. To half a pint of boiling milk add one or two wine-glassfuls of Sherry or Madeira. Separate the curd by straining through a fine sieve or piece of muslin. Sweeten the whey with refined sugar.

The following is a useful and agreeable demulcent drink :

251. R. Extracti sarsaparillæ fluidi,
          Syrupi scillæ,   aa. f.ʒjss.   M.
      A teaspoonful in a teacupful of barley-water, to be frequently taken during the day.

Then, after a brisk purgative, either of the following may be administered :

252. R. Syrupi scillæ,   f.ʒvj.
          Spiritûs ætheris nitrosi,
          Tincturæ hyoscyami,   aa. f.ʒiij.
          Infusi rosæ acidi,   f.ʒjss.   M.
      A tablespoonful every six hours.

253. R. Potassæ nitratis,   Əij.
          Vini antimonii,   f.ʒj.
          Liquoris ammoniæ acetatis, f.ʒss.
          Aquæ camphoræ,   q.s. ad f.ʒiij.   M.
      A tablespoonful every four hours.

6

If there be any depression, stimulating expectorants, such as the following, must be ordered:

254. R. Ammoniæ carbonatis,   ʒss.
          Spiritûs ætheris sulphurici, f.ʒiij.
          Tincturæ scillæ,   f.ʒjss.
          Tinct. opii comphoratæ,   f.ʒij-iv.
          Tinct. lavendulæ compositæ, f.ʒvj.
          Infusi senegæ,   q.s ad f.ʒviij.   M.

Two tablespoonfuls every four hours.

255. R. Spiritûs ammoniæ aromatici, f.ʒss.
          Spiritûs chloroformi,   f.ʒiij.
          Tincturæ aconiti,   f.ʒss.
          Tincturæ senegæ,   f.ʒvj.
          Aquæ camphoræ,   q.s. ad f.ʒiij.   M.

A tablespoonful every six hours.

Gentle counter-irritation to the front of the chest by dry cupping, turpentine stupes or sinapisms will prove useful. Should the phlegm appear to accumulate in the bronchial tubes the following emetic will readily remove it:

256. R. Antimonii et potassæ tart., gr.j.-ij
          Vini ipecacuanhæ,   f.ʒij.   M.

For one dose, to be added to a wineglassful of water and its action aided by the free administration of warm water *

CHRONIC BRONCHITIS.

WILLIAM AITKEN, M. D., EDIN.

In chronic cases of bronchitis, especially in patients who have made considerable progress in the journey of life, remedies which tend to invigorate the general system are indicated. Besides the selection of a beneficial climate and the use of nutritious, easily digested food, stimulating embrocations are useful. The following liniment, employed at the Meath Hospital at Dublin, is highly recommended for this purpose:

*Practice of Medicine: p. 341.

257. ℞. Spiritûs terebinthinæ, f.ʒiij.
   Acidi acetici, f.ʒss.
   Vitelli ovi, j.
   Aquæ rosæ, f.ʒijss.
   Olei limonis, f.ʒj.   M.

To be rubbed, morning and evening, not only over the chest before and behind, but along the sides of the neck. It generally reddens the skin and produces small pimples.

Of the fetid gums, ammoniac in particular is a useful remedy. From the following combination (formula of Prof. Easton of Glasgow) decided beneficial results are obtained:

258. ℞. Ammoniaci, ʒij.
   Acidi nitrici diluti, f.ʒij.
   Aquæ, f.ʒxij.   M.
Two tablespoonfuls in gruel, ter die.

It is often advantageous to astringent remedies, give as

259. ℞. Acidi tannici, Эj-ʒj.
For xx. pills. One ter die. Or,

260. ℞. Olei cubebi, gtt.x.
For one dose three or four times a day on a piece of sugar.

In the protracted bronchitic affections of the aged, diuretics are of great service. The following formulæ are recommended by our author and by Drs. MACLACHLAN and STOKES, as well suited in a variety of cases of *senile chronic catarrh:*

261. ℞. Potassæ nitratis, gr. xxxvj.
   Tincturæ conii, f.ʒss.
   Spiritûs ætheris nitrosi,
   Oxymellis scillæ, aa f.ʒvj.
   Decocti senegæ, q.s. ad f.ʒvj.   M.
A tablespoonful ter die.

262. ℞. Potassæ acetatis, ʒss.
   Aceti scillæ,
   Spiritûs ætheris nitrosi, aa f.ʒvj.
   Tinct. opii camphoratæ, f.ʒss.
   Liquoris ammon. acetatis, f.ʒiijss.
   Syrupi aurantii corticis, f.ʒvj.   M.
A tablespoonful ter die.*

---

* Science and Practice of Medicine, Am. Ed., vol. ii., p. 703.

INHALATION.

DR. BEIGEL, OF LONDON.

263. R. Pulveris aluminis,   gr. ij.
          Tincturæ opii,   ♏x.
          Aquæ destillatæ,   f.℥j.   M.

For inhalation, by means of atomizer, night and morning,
in obstinate cases of chronic bronchitis.

The following inhalation also affords relief in severe cases :

264. R. Acidi tannici,   gr. iij.
          Extracti hyoscyami,   gr. ij
          Aquæ destillatæ,   f.℥j.   M.

To be inhaled night and morning.   On account of the taste,
ferri sulphatis (gr. iv. to f.℥j) may be substituted for the
tannic acid.*

J. SOLIS COHEN, M. D., PHILADELPHIA.

265. R. Liquoris iodinii compositi, gtt. x-xl.
          Aquæ destillæ,   f.℥j.   M.

For inhalation by means of atomizer two and three times a
day.   If there co-exist pain, the narcotics and sedatives
may be added to the inhalation, but preferably in minute
quantity.†

DR. VAN DER CORPUT.

266. R. Morphiæ muriatis,   gr. j.
          Ammoniaci,   ℨss.
          Extracti scillæ,   gr. xv-xxv.   M.

For xx pills—take from two to four in the course of the day,
in chronic bronchitis and bronchorrhœa.‡

J. M. DA COSTA, PHILADELPHIA.

267. R. Ammoniæ muriatis,   ℨij.
          Misturæ glycyrrhizæ comp.,   f.℥iij.   M.

A dessertspoonful three times a day.

Muriate of ammonia in order to be effective should be given
in ten grain doses.   In the bronchitis of patients affected with
phthisis, it may be combined as follows :

* Cohen on Inhalation, p. 149.
† On Inhalation, p. 149.
‡ Formulaire Raisonné des Médicaments Nouveaux, p. 449.

268. R. Ammoniæ muriatis, ℥ss.
Morphiæ muriatis, gr. j.
Ext. pruni virg. fluidi, f.℥iij. M.

A teaspoonful three or four times a day.

269. R. Ammoniæ muriatis, ʒij.
Potassæ chloratis, ʒj.
Tincturæ hyoscyami, f.℥ss.
Ext. pruni virg. fluidi, f.℥ijss. M.

A dessertspoonful, ter die, when there is a tenacious secretion.

Chlorate of potash thins the secretion and promotes expectoration; it is useful in both acute and chronic bronchitis.

270. R. Potassæ chloratis, ʒij.
Tincturæ scillæ, f.℥ss.
Misturæ glycyrrh. comp., f.℥ijss. M.

A dessertspoonful three or four times a day when there are dry râles.

271. R. Vini picis liquidæ. f℥iij.

A dessertspoonful ter die in cases of bronchial catarrh, together with

272. R. Plumbi acetatis, Əij.
Extracti glycyrrhizæ, q. s. M.

For xx pills.

One three times a day.

273. R. Potassæ carbonatis, ℥ss.
Vini ipecacuanhæ, f.ʒij.
Tincturæ opii camphoratæ,
Syrupi tolutanus, aa f.℥jss. M.

A teaspoonful three times a day as an alkaline expectorant mixture to thin the secretion.

Or, the following may be used when a tonic is also indicated:

274. R. Sodæ carbonatis, ʒj.
Ammoniæ muriatis, ʒij.
Extracti gentianæ, fluidi, f.ʒvj.
Extracti hyoscyami fluidi, f.ʒij.
Syrupi tolutanus, f.℥ij M.

A dessertspoonful three times a day.

INHALATION.

275. R. Pulveris alumenis,   gr. viij.
          Extracti conii fluidi,   gtt. vj.
          Aquæ destillatæ,   f.ʒj.   M.
      For atomization. The alum to be gradually increased
      to gr. xx. to f.ʒj.

PROF. A. P. DUTCHER, M. D., CLEVELAND, OHIO.

As *local remedies* the inhalation of the vapor of hops,
iodine, chloroform, tar, extract of conium and belladonna are
at times very useful. The *vapor of iodine* should be watched
with the greatest attention. If it increases dyspnœa and
produces an unusual feeling of heat and distress in the bronchial region, it should be immediately discontinued, for it
will produce congestion in the smaller bronchia—which may
extend to the air cells and ultimately end in pneumonia.

The *constitutional treatment* must vary with the wants of
each particular case. Tonics, such as iron, gentian and
quinine, with minute doses of mercury, are commonly useful
to improve the state of the secretions and functions generally.
When the expectoration is profuse, with much febrile excitement, the mineral acids and metallic astringents are useful.
When the urine is loaded with the oxalate of lime and the
bronchial trouble appears to depend upon indigestion, from
some defect in the functions of primary assimilation, the
nitro-muriatic acid, administered according to the following
formula, will sometimes produce a wonderful change for the
better, in cases that have resisted every other form of medication :

276. R. Acidi nitrici,   f.ʒj.
          Acidi muriatici,   f.ʒij.
          Morphiæ sulphatis,   gr. ij.
          Tr. cinchoniæ compositæ,   f.ʒiv.   M.
      A teaspoonful ter die before each meal.

In mild cases of chronic bronchitis the iodide of potash is a

most valuable remedy ; it seems to restrain low degrees of
inflammation affecting the fibrous part of the tubes, and coun-
teracts the process of induration to which they tend. In bad
cases of the disease, where there is no tendency to tuberculosis,
our author is in the habit of prescribing the iodide of potash
and corrosive chloride of mercury, thus :

> 277. R. Hydrarg. chlor. corrosivi,   gr. j.
> Potassii iodidi,   ℥iij.
> Extracti lobeliæ fluidi,   f.℥j.
> Syrupi simplicis,   f.℥v.   M.
> A teaspoonful three times a day, after each meal.

Where there is a tendency to tuberculosis the mercury
should be omitted.

The following is a useful combination where it can be borne
by the stomach, to allay cough and restrain expectoration :

> 278. R. Copaibæ,   f.℥ij.
> Tincturæ cubebæ,   f.℥j.
> Morphiæ sulphatis,   gr.iv.
> Syrupi simplicis,   f.℥j.   M.
> A teaspoonful three times a day.

Ipecacuanha as an expectorant, in ordinary cases of chronic
bronchitis, may be given in doses of a grain or two of the
powder, or twenty to thirty drops of the wine, repeated several times a day, or combined with other agents, according to
the following :

> 279. R. Vini ipecacuanhæ,
> Syrupi scillæ   aa f.℥j.
> Tincturæ digitalis,
> Spiritûs ætheris nitrosi,   aa f.℥ss.
> Tincturæ opii camphoratæ,   f.℥ij.   M.
> A teaspoonful three times a day.

In the chronic bronchitis of aged people, where, from alterations in the structure of the tubes, a cure cannot be expected, the disease may often be palliated and the patient ren-

dered very comfortable by the use of the following combination, which is one of our author's favorite prescriptions :

280. R. Potassii ferrocyanureti,   ℥iv.
          Morphiæ sulphatis,   gr. v.
          Tincturæ colchici radicis,
          Syrupi scillæ,   aa f.℥ss.
          Aquæ destillatæ,   f.℥iv.   M.

A teaspoonful three or four times a day.

When expectoration is viscid, alkalies are useful; and when the vital powers are feeble Dr. D. is in the habit of prescribing, in connection with other treatment, the following :

281. R. Ferri pyrophosphatis,
          Quiniæ sulphatis,   aa ℨj.
          Strychniæ,   gr.j.
          Extracti hyoscyami,   gr.xxx.   M.

For lx pills. Two to be taken three or four times a day after each meal.[*]

282. R. Potassii iodidi,   ℥iij.
          Hydrarg. chlor. corrosivi,   gr.ij.
          Extracti lobeliæ fluidi,   f.℥j.
          Syrupi stillingiæ compositi,
          Syrupi phellandrii aquatici
              compositi,   aa. f.℥vijss. M.

One-half an ounce, three times a day, before each meal; and the chest over the bronchial regions to be painted most thoroughly every night on retiring to rest, with

283. R. Ioidinii,   ℨj.
          Potassi. iodidi,   ℨij.
          Aquæ,   f.℥iv.   .   M.

INHALATION.

284. R. Iodinii,   gr.xx.
          Chloroformi,   f.℥j.   M.

Forty drops to be inhaled, every night before going to bed. After which, to quiet cough and secure good rest at night take one of the following :

*Philadelphia *Medical and Surgical Reporter* for October 19th, 1867 p. 329.

285. ℞.   Quiniæ sulphatis,
           Extracti hyoscyami,   aa ʒss.
           Morphiæ sulphatis,   gr.iij.   M.
      For xv pills.*

INHALATION.

DR. FREDERICK FIEBER, OF VIENNA.

286. ℞.   Zinci sulphatis,   gr. v.
           Aquæ destillatæ,   f.ʒj.   M.

Of marked benefit in a case of chronic bronchitis, of the
variety simulating consumption, of twenty-five years' standing.†

DR. E. HEADLAM GREENHOW, F. R. C. P., ETC.

287. ℞.   Vini ipecacuanhæ.
           Acidi nitro-muriatici dil., aa ℳx.
           Tr. hyoscyami,   ℳxx.
           Tr. gentianæ compositæ,   f.ʒss.
           Aquæ,   ad f.ʒij.   M.
      For one dose.   To be taken in water three times a day, in
      chronic bronchitis.

In almost all cases of chronic bronchitis a time arrives when
expectorants cease to be useful. The expectoration has become of the nature of an habitual flow from the bronchial
membrane.   Treatment of a tonic character is then required.
In these cases Dr. Greenhow has long been accustomed to
prescribe with great advantage the mineral acids, especially
the nitro-muriatic, in combination with a vegetable bitter,
as above.   In chronic cases attended by very copious expectoration, he adds to each dose of the above mixture twenty
minims of the tincture of larch, which has the effect not only
of lessening the expectoration, and with it the cough and
dyspnœa, but also apparently of restoring the debilitated mem-

*Philadelphia *Medical and Surgical Reporter* for October 12, 1867.
† Cohen on Inhalation. p. 149.

brane to a more healthy tone, and of rendering patients less liable to catarrhal attacks at every change of the weather or season.*

INHALATION,

DR. JOHN FORSYTH MEIGS, PHILADELPHIA.

288. R. Acidi carbolici fluidi, ℥x-xv.
       Aquæ, Oss. M.
To be placed in an ordinary inhaling bottle, and used three or four times a day, in bronchial catarrh, offensive secretion from bronchial tubes, etc.

PROF. ALFRED STILLÉ, M. D., PHILADELPHIA.

289. R. Acidi carbolici fluidi, gtt.xv.xx.
       Tincturæ conii, f.℥j-ij.
       Aquæ destillatæ, Oij. M.

This solution should be inhaled by means of an atomizer; useful both in the simple form of chronic bronchitis and in that which usually complicates the advanced stages of phthisis. It diminishes the secretion and thereby lessens the waste of substance and the exhaustion occasioned by repeated and urgent coughing.

ELECTUARY AGAINST CHRONIC BRONCHITIS.

L. PARISEL, OF PARIS.

290. R. Pulv. cinchonæ calisayæ,
       Sulphuris loti, aa ℥ss.
       Syrupi althæa, q. s. M.
One teaspoonful every four hours.

It modifies the morbid bronchial secretion, facilitates expectoration, and regulates the digestive functions.†

---

*On Chronic Bronchitis, Am. Ed., 1869, p. 52.
†Annuaire Pharmaceutique, par L. Parisel, 1867, p. 229.

THOMAS HAWKES TANNER, M. D,

291. R. Syrupi scillæ, f.ʒvj.
Acidi nitrici diluti, f.ʒj.
Tincturæ hyoscyami, f.ʒiij.
Spiritûs chloroformi, f.ʒvj.
Infusi cinchonæ flavæ, q. s. ad. f.ʒvj. M.

Two tablespoonfuls twice or thrice daily, in chronic catarrh, with debility and restlessness.

292. R. Syrupi scillæ, f.ʒvj.
Spiritûs ammoniæ aromatici, f.ʒiij.
Morphiæ muriatis, gr. ½.
Infusi serpentariæ, q. s. ad f.ʒvj. M.

Two tablespoonfuls two or three times a day, in chronic catarrh.

293. R. Pilulæ scillæ compositæ,
Extracti conii, aa ʒss. M.
For xii pills.

Two to be taken every night at bed time.

In chronic catarrh when opium is objectionable.

294. R. Tincturæ scillæ, f.ʒij.
Tincturæ stramonii, f.ʒjss.
Infusi dulcamaræ, q. s. ad f.ʒvj. M.

Two tablespoonfuls ter die.

In chronic catarrh, especially when the secretions of the skin and kidneys are deficient.

INHALATION.

295. R. Olei terebinthinæ, f.ʒj.
Aquæ calidæ, t.ʒv. M.

In chronic bronchitis with excessive secretion.

296. R. Creasoti. m.xxx.
Aquæ bullientis, f.ʒviij. M.

In chronic catarrh.

EMPHYSEMA.

WILLIAM AITKEN, M. D., EDIN.

Little can be done, apart from the treatment of the bronchial congestion. If bronchial spasm prevail, the following may give relief:

297. R. Spiritûs ætheris comp.,   f ℨiv–viij.
        Aquæ camphoræ.   f.℥iv.   M.
A tablespoonful.*

THOMAS KING CHAMBERS, M. D., ETC., LONDON.

298. R. Tincturæ ferri chloridi,   f.ℨijss.
       Tincturæ lobeliæ æthereæ,   f.ℨij.
       Aquæ camphoræ,   f.℥iv.   M.
A tablespoonful, in water, ter die.

The object of the iron is to try and restore its full vital powers to the creative arterial blood, so that it may renew the pulmonary membrane, that it may form healthy elastic tissue, instead of the imperfectly elastic degenerated tissue. The lobelia is ordered as a substitute for a more powerful medicinal agent, tobacco. Nothing calms the distressing asthma so well as a few whiffs of strong Virginia. Like tobacco, lobelia is a very variable article; there seems to be as much difference between one specimen and another, as between the mildest cigarette and the strongest shag. The strongest sort should be used, paying for it the best price at the best shop, so that it may be given in moderate and graduated doses.†

MEREDITH CLYMER, M. D., ETC., NEW YORK.

For the relief of the asthmatic fits of emphysema, nothing is so sure as a full dose of opium with sulphuric or chloric ether:

299. R. Tincturæ opii,   f.ℨj.
       Ætheris sulphurici,   f.ℨij.   M.
Sixty drops every twenty minutes.

To each dose may be added
Tincturæ lobeliæ æthereæ, gtt. xx.

---

*Science and Practice of Medicine. Am. Ed., 1868, vol. ii., p. 717.
† The Renewal of Life. Am. Ed., p. 292.

J. M. DA COSTA, M. D., PHILADELPHIA.

300. R. Potassæ chloratis, ʒjss.
Tincturæ belladonnæ, f.ʒjss.
Syr. pruni virginiani fluidi,
Tinct. cinchonæ comp., aa f.ʒij. M.

A dessert-spoonful four times a day, in emphysema with chronic bronchitis and loss of appetite. Also, dry cups applied to the chest morning and evening.

In the treatment of emphysema, strychnia and nux vomica are of no benefit. Chlorate of potash in large doses is of service, so also is the iodide of potassium. Care should be taken to prevent the emphysematous patient from having attacks of bronchitis, which aggravate the affection.

In the bronchial complications of emphysema the following formula will be found useful:

301. R. Tincturæ lobeliæ, f.ʒj.
Syrupi scillæ,
Syrupi tolutanus, aa f.ʒss.
Misturæ glycyrrhizæ comp., f.ʒiv. M.

A dessert-spoonful every three hours; with counter-irritation to the chest with the following:

302. R. Acidi acetici,
Olei terebinthinæ,
Linimenti saponis, aa fʒij. M.

To be rubbed on night and morning.

303. R. Potassii iodidi, ʒij.
Syrupi ipecacuanhæ,
Tincturæ scillæ, aa f.ʒss.
Syrupi simplicis, f.ʒij. M.

A teaspoonful ter die.

Together with counter-irritation to the chest:

304. R. Chloroformi, f.ʒss.
Linimenti ammoniæ,
Linimenti saponis, aa f.ʒjss. M.

To be rubbed on morning and evening, and to be placed on flannel, and allowed to remain against the skin for fifteen minutes.

305. ℞. Potassii iodidi,   ℥ij.
       Extracti senegæ fluidi,   f.ʒj.
       Syrupi pruni virg. fluidi,   f.℥ij.   M.
A teaspoonful ter die.

306. ℞. Tincturæ lobeliæ,   f.℥ss.
       Extracti valerianæ fluidi,   f.℥j.
       Spiritûs ætheris compositi,   f.℥j.
       Potassæ chloratis,   ℈iv.
       Syrupi tolutanus,   f.℥j.
       Aquæ,   f.℥ss.   M.
A dessertspoonful, in water, four times a day, or oftener if
the oppression be great

Also, the following :

307. ℞. Chloroformi,   f.℥ss.
       Olei terebinthinæ,   f.℥j.
       Spiritûs rosmarini,   f.℥jss.   M.
To be rubbed on morning and evening.

308. ℞. Atropiæ sulphatis,   gr. 1-60.
In granules ter die.

The treatment in cases of emphysema should be a double
one, to modify the bronchial trouble which keeps up the affection, and to alleviate the difficulty of breathing, which may
at times rise into paroxysms of attacks of asthma. In point of
radical treatment, there are no means which will cause the
distended air vessels to resume their natural size. As a matter of absolute experience our author has thought that he has
seen in cases of pure emphysema, not associated with bronchial
symptoms, a long course of iodide of potassium (three to five
grains ter die, for several months,) favorably influence the
disease, the respiratory murmur becoming fuller and freer,
and the prominence of the chest walls less visible. Good results are also obtained from persistent counter-irritation.
Small fly blisters at various portions of the chest are serviceable. A number of cases are reported in which the constant employment of the continuous galvanic current has led
to a diminution in the size of the chest.

E. HEADLAM GREENHOW, M. D., F. R. C. P., ETC., LONDON.

309. ℞. Potassii iodidi, gr. xxxvj.
Ammoniæ carbonatis, ʒj.
Tincturæ scillæ,
Tincturæ hyoscyami, aa f.℥ss.
Aquæ camphoræ, f.ʒv. M.

A tablespoonful ter die. Also, direct the patient to smoke
a stramonium cigarette so soon as he feels the commencement of an asthmatic attack.*

DR. FELIX VON NIEMEYER, PROFESSOR, UNIVERSITY OF TU-
BINGEN.

The symptomatic indications in emphysema (our author
considers we are totally unable to fulfil the indications of the
disease itself, the nutritive alterations upon which it depends
being irreparable) are first, the proper treatment of the bronchial catarrh, which almost always accompanies this affection,
and greatly adds to the distress of the patient. Habitual
wearing of flannel next the skin, stimulants to the chest, warm
baths of water or vapor, and the alkaline muriatic mineral
springs (especially the thermal springs of Ems) are often of
signal benefit.

The next symptomatic indication is to moderate the habitual shortness of breath and the attacks of severe dyspnœa.
Sending the patient during the summer to the pine wood region, and particularly to places where there is a heavy fall of
dew, will allay the persistent oppression of the chest. The
inhalation of compressed air, for the same reason, is an excellent palliative. To avert the asthmatic attacks, a strict diet,
the avoidance of food likely to induce flatulence, light suppers
and the keeping of the bowels open are required.

During the attacks of emphysematous asthma the narcotics
should be used with caution, unless called for by bronchial

* On Chronic Bronchitis. Am. Ed. 1869, p. 152.

spasm. The more suitable remedies (beside the emetics, which are very appropriate,) are the stimulants, camphor, musk, benzine, and

310. R. Port wine, f.℥j-jss.
Every three hours.

When these fail, use

311. R. Olei terebinthinæ, f.℥j-iv.
Aquæ menthæ piperitæ, f.℥iv.
Sacchari albi,
Pulveris acaciæ, aa ℥j. M.
A tablespoonful every three hours.

For the dropsy complicating the affection, vigorous diaphoresis will give excellent results whenever it depends upon a capillary bronchitis. Later in the disease, when it arises from failure of the heart to compensate for the circulatory derangement of the lungs, it may be relieved for a time by

312. R. Pulveris digitalis, Əss-j.
Aquæ, f.℥vj. M.
For an infusion.
A tablespoonful ter die.

Where digitalis fails, squills may be employed.

313. R. Aceti scillæ, f.℥j.
Aquæ destillatæ, f.℥vj.
Potassæ carb., q. s. ad sat. M.
A tablespoonful every two hours.*

THOMAS HAWKES TANNER, M. D., F. L. S. ETC., LONDON.

314. R. Spiritûs ætheris compositi, f.℥jss.
Spiritûs ammoniæ aromatici, f.℥ij.
Tincturæ opii camphoratæ, f.℥jss.
Aquæ camphoræ, q. s. ad f.℥iv. M.
Two tablespoonfuls every half hour until the spasm is relieved.

---

* Text Book of Practical Medicine, Am. Ed., 1869, vol. i., p. 125.

315. ℞. Sambucii radicis, ℥ss.
Spiritûs ætheris compositi, f.℥iv. M.
Maccrate in a stopped bottle for seven days, and then filter.
Dose, ℳ xx-xxx.*

PHTHISIS PULMONALIS AND SCROFULA.

WILLIAM AITKEN, M. D., EDINBURGH, ETC.

316. ℞. Olei morrhuæ, f.℥jss.
Olei creasoti, gtt iv.
Pulveris tragacanthæ,
Pulveris acaciæ,
Pulveris amyli, aa ℈j.
Sacchari albi, ℥j.
Aquæ anisi, f.℥ivss. M.
Take two tablespoonfuls three times a day.

This is recommended as making a palatable mixture. The creasote is said to render the stomach more tolerant of the remedy.

Besides cod-liver oil, other animal fats and oils, where they can be taken and assimilated, are sure to be followed with benefit. Hence milk, rich in fatty matters, such as asses' milk, and milk drawn from cows at a short interval after the greater part of their milk has been withdrawn, are found to be followed by improvement, where they are persevered in and are assimilated. So, also, with cream and butter. Dr. BENNETT instances the partial success occasionally of caviar, bacon, pork, mutton chops and the marrow of bones of oxen; while Dr. THOMPSON instances the good effects he has obtained from the use of oil from the foot of the young heifer (neat's foot oil.)

But medicine is utterly powerless and useless, unless hygienic means are carried out to the uttermost. They may be enumerated as follows:

---

*Practice of Medicine, Am. Ed., p. 354.
7

I. A constant supply of pure and fresh air for respiration.

II. Active exercise in the open air. *The risk is in staying in the house,* and not in going out of it.

III. It is important to secure for the patient a uniform, sheltered, temperate, and mild climate to live in, with a temperature about 60° and a range of not more than 10° or 15°; where, also, the soil is dry and the drinking-water pure and not hard.

IV. The dress of the scrofulous patient ought to be of such a kind as to equalize and retain the temperature of the body. Waterproof coats, boots, and shoes are to be condemned. Flannel ought to be invariably worn next the skin in all seasons.

V. The hours of rest should extend from sunset to sunrise.

VI. In-door or sedentary occupation should be suspended; but out-door employment in the fresh air, even in the midst of snow, has been and may be advantageous.

VII. Cleanliness of body is a special point to be attended to.

VIII. Marriage of consumptive females, for the sake of arresting the disease by pregnancy, is morally wrong and physically mischievous.

IX. The medicinal treatment must be adapted to the site of the local deposits and the general nature of the particular case.

To promote and preserve an appetite for food should be constantly kept in view as one of the great objects of treatment. One of the best tonics which can be employed is that proposed by Professor EASTON, of Glasgow; it should be given in very small doses at first, and be followed by the use of the cleanest and most agreeable kind of cod-liver oil. The following chalybeate very rarely disagrees :

317. ℞. Vini ferri,   f.ʒij.
A teaspoonful ter die.

318. ℞. Pilulæ saponis comp.,
          Pilulæ scillæ comp.,   aa. ℈j.   M.
For viij pills. One at bed time to procure sleep. Two may
  be given if the cough is troublesome. Or,

319. ℞. Pulv. ipecacuanhæ comp.,   ℈iv.
          Tincturæ scillæ,
          Tincturæ tolutanus,   aa f.ʒij.
          Misturæ acaciæ,   f.ʒjss.
          Aquæ,   q. s. ad f.ʒiij.   M.
A dessert-spoonful at night to quiet the cough.*

MEREDITH CLYMER, M. D., ETC., NEW YORK.

Counter-irritation to the chest walls, in the earlier stages,
before there is much loss of strength is undoubtely beneficial;
but later is weakening and annoying. Croton oil liniment is
chiefly used for this purpose. A prompt and not too severe
application is the following ointment (recommended by Dr.
Fuller):

320. ℞. Hydrargyri chloridi mitis, gr. viij.
          Iodinii,   ʒss
          Alcoholis,   f.ʒjss.
          Unguenti adipis,   ʒj.   M.
Rub in a portion over the affected lung, morning and evening, until a pustular eruption comes out.†

J. WARING CURRAN, L. K. & Q. C. P., ETC.

321. ℞. Zinci oxidi,   gr. ij.
          Extracti conii,   gr. j.   M.
For one pill, to be taken three times a day. The quantity
  of the oxide of zinc is gradually to be increased.

In the latter stages of phthisis, where profuse sweating and
colliquative diarrhœa harass the patient and rapidly lower
the vital capacity, this combination is very effective. It is of
great value also in the earlier stages of the disease. It seems

*Science and Practice of Medicine, vol. ii., Am. Ed., p. 794.
†Aitken's Science and Practice of Medicine, Am. Ed., vol. ii., p. 797.

to steady the nervous system and act as a sedative to the wandering pains.

J. M. DA COSTA, M. D., PHILADELPHIA.

322. R. Morphiæ acetatis,   gr. ij.
      Potassii cyanidi,   gr. j.
      Acidi acetici,   f.ʒj.
      Ext. pruni virginianæ fl.,
      Misturæ acaciæ,   aa f.ʒij.   M.

A teaspoonful four or six times a day, as a sedative mixture for the cough of phthisis.

323. R. Liquoris morphiæ sulphatis, f.ʒj.
      Ext. pruni virginianæ fl.,   f.ʒij.
      Acidi sulphurici diluti,   f.ʒij.   M.

A teaspoonful three or four times a day when night sweats and cough are troublesome.

324. R. Syrupi hypophosphitis,   f.ʒiij.

A teaspoonful ter die, after meals.

INHALATION.

325. R. Extracti opii,   gr. ss.
      Aquæ,   f.ʒj.

For one inhalation, twice a day, by means of any form of steam atomizer throwing a fine spray. In the irritative cough of phthisis, causing gastric irritability.

326. R. Tincturæ iodinii compositæ, ℳ x.
      Aquæ,   f.ʒj.   M.

   For atomization.

327. R. Tincturæ ferri chloridi,   f.ʒj.
      Acidi muriatici diluti,   f.ʒij.   M.

Twenty-five drops in sweetened water drawn through a tube before meals. In the treatment of tubercles in the lungs, complicated with tubercular diarrhœa and impairment of digestion.

Also the following:

SUPPOSITORY.

328. R. Extracti opii,   ℈j.
      Plumbi acetatis,   ℈ij.   M.

Make into xx suppositories. One to be introduced morning and evening.

329. R. Syrupi hypophosphitis,
　　　　 Ext. pruni virgin. fluidi, aa f.℥ij.　　　M.
A dessertspoonful ter die.

330. R. Olei morrhuæ,　　　　　　　　f.℥j.
　　　　 Aquæ menthæ piperitæ,　　　　f.℥ss.
　　　　 Tincturæ aurantii corticis,　　 f℥ss.
　　　　 Misturæ acaciæ,　　　　　　　f℥iijss.
　　　　 Olei gaultheriæ,　　　　　　　℔x.　　　M.
A dessertspoonful three times a day. This formula disguises
somewhat the taste of the cod-liver oil, or,

J. M. DA COSTA, M. D., PHILADELPIA.

331. R. Olei morrhuæ,　　　　　　　f.℥ss.
For one dose ; to be taken three times a day in *carbonic acid
water*.

Place in a tumbler a small amount of any preferred syrup
(orgeat or sarsaparilla is the best adapted to disguise the taste
of the oil) and fill up with carbonic acid water, from a bottle
furnished with a syphon for table use ; then, while it is still
foaming, put in a tablespoonful of the oil. It is astonishing
how perfectly the taste is concealed in this manner. Other
modes of taking the oil are floating on ice water, in lemon
juice, and in the froth of porter. Less than a tablespoonful
is not worth taking. The best time for its administration is
between meals, after the process of digestion is pretty well finished. Persons are exceptionally met with who take it in
preference just before meals, and thus avoid the disagreeable
eructations. Nobody likes to take it the instant after meals.

332. R. Acidi arseniosi,　　　　　　gr.j.
　　　　 Feri lactatis,　　　　　　　℥ss.
　　　　 Syrupi,　　　　　　　　　　q. s.　　M.
For xxx pills ; one ter die.

Arsenic is an agent which may frequently be employed with
advantage in cases of slow consumption.

It may be given as above or as follows :

333. ℞. Liquoris potassæ arsenitis, f.ʒij
Ext. pruni virgin. fluidi, f.ʒiij.   M.
A teaspoonful ter die.

334. ℞. Calcis hypophosphitis, ʒss.
Sodæ hypophosphitis, ʒij.
Misturæ acaciæ, f.ʒiij.   M.

A teaspoonful ter die, with plenty of cream, eggs, etc., about three ounces of whisky daily and F. 322 for the irritative cough. In a case in which cavities had formed in the lungs.

335. ℞. Quiniæ sulphatis, Əij.
Acidi tannici, Əj.
Extracti gentianæ, q. s.   M.
For xx pills. One ter die, to reduce night sweats.

CONSUMPTION HOSPTAL, LONDON.

336. ℞. Morphiæ muriatis, gr.ss.
Acidi hydrocyanici diluti, ℳxv.
Acidi muriatici diluti, ℳijss.
Oxymellis scillæ, f.ʒss.
Aquæ, q. s. ad f.ʒj.   M.
Dose—One to two drachms.*

BALTHAZAR W. FOSTER, M. D., M. R. C. P., PROFESSOR OF MEDICINE IN QUEEN'S COLLEGE, ETC., LONDON.

337. ℞. Æther, f.ʒijss.
Olei morrhuæ, f.ʒiv.   M.
Sig.—Dessertspoonful ter die, before meals.

This is a new physiological attempt to introduce cod-liver oil into the system, by overcoming the difficulty of assimilating fat, which is developed to the greatest extent at the very stages of the disease in which perfect assimilation is most needed. To pour oil into a patient's stomach, without at the same time taking measures to ensure its digestion, is a crude kind of therapeutics. Experimental physiology has taught us that the only fluids in the body which have the power of acting upon fat, so as to render it fit for absorption, are the

*Squire's Pharmacopœia of the London Hospitals, 2d Ed., 1869, p. 170.

secretions of the pancreas and the duodenal glands. Ether has the power of stimulating the glands to renew their healthy action, and places the fatty matter in a state of fine division with their abundant secretion. It also masks the unpleasant taste of the oil.

M. FURTER, OF MONTPELLIER.

Our author reported sometime ago, before the Paris Academy, his treatment of phthisis, as employed with good results at the clinic in Montpellier, under his charge. It consists in the use of raw mutton or beef, given in conjunction with strongly diluted alcohol, in small doses.

338. R. Take some finely minced raw mutton or beef and
roll it up in sugar or in saccharine electuary. Give
in teaspoonful doses to the amount of 3-10 ounces
per day.

339. R. Alcoholis,   f.℥j.
Syrupi,   f.℥iij.   M.
A teaspoonful dose every hour. The dose and frequency
of administration are to be modified by the patient's in-
dividuality.

MM. MONTARD-MARTIN AND HÉRARD, PARIS.

340. R   Acidi arseniosi,   gr.j.
In granules lx. div.
Seven or eight a day of these granules to be taken at first,
the dose to be speedily carried up to ten or fifteen.
Never more than two should be given at a time, and they
should be administered as often as possible before meals.
The treatment should be suspended from time to time.

Researches have shown the entire harmlessness of arsenic, when properly employed, as well as its undeniable efficacy in certain forms of tubercular phthisis. Almost all the patients after a few days, treatment exhibited a marked improvement in their general condition. The appetite improves, the strength returns, the complexion is clearer, and the eye is more animated ; and at the end of three weeks or a month flesh begins to be gained. The local malady undergoes less change, but even this is sometimes sensibly modified. The most favorable cases are those in which there is no acute fever or serious digestive disturbance. M. LOLLOIT, the most recent observer, finds that the daily administration of one-tenth of a grain produces a diminution of temperature, and a very notable diminution in the amount of urea.

DR. FELIX VON NIEMEYER, PROFESSOR OF PATHOLOGY AND THERAPEUTICS, DIRECTOR OF THE MEDICAL CLINIC OF THE UNIVERSITY OF TUBINGEN.

Cod-liver oil has a special and well-merited reputation as a remedy against scrofula, and there are plenty of instances where it has been of good service. On the other hand, perhaps no remedy has ever been so much abused as this one. Whoever supposes that the mere presence of a thick nose, a sore upper lip, or a bunch of enlarged cervical glands, affords sufficient grounds for the prescription of this medicine, will often fail to benefit his patient, and sometimes will do him harm. Daily experience teaches, however, that such is the general belief, and that he who seeks to combat it does not merely fight a windmill. Let any one ask a patient whose scrofula has outlasted his childhood, and who has passed again and again from one practitioner to another, how often he has had cod-liver oil prescribed for him since the time of its first failure during childhood; how many months or years he has taken it; and how much the whole aggregate quantity would amount to, and he would be surprised at the answer. Nevertheless, in all probability, the next physician whom the patient consults will prescribe it again. A most serviceable means of distinction between the cases in which cod-liver oil is indicated and those in which nothing is to be expected from it, is afforded by the symptoms of the torpid and erethitic forms of scrofula. When the patient's slender frame, the lack of fat beneath his skin, and his accelerated pulse warrant the belief that his nervous system is in a state of over-activity, cod-liver oil is generally of most signal benefit. Under its use the plumpness of the body increases, while the general susceptibility of the system, and the diseases consequent upon it. subside. These are the cases to which this article owes its name as an anti-scrofulous remedy. But if the patient be clumsy and thick-set; if the nose and upper lip be enlarged, and the adipose layer over the rest of the body strongly developed; if the action of the heart be retarded rather than accelerated; if the irritability of the nervous system seem unusually obtuse; in short, should there be reason to suppose that the waste of the system is diminished rather than increased, we cannot hope to relieve the disease by means of the oil. Nevertheless, it is precisely this class of patients who in vain have taken such enormous quantities of it in the course of their lives. Besides the oil, and as a corroborant of its effects, so to speak, articles containing a little tannin, such as parched acorns, "acorn coffee," and home-made infusion of walnut leaves are very often prescribed. Such a practice is greatly to be commended whenever there is a chronic catarrh of the intestines, embracing the digestion and the absorption of chyle, and where apprehensions are entertained that the oil may aggravate the intestinal disorder. In order to make children take the acorn coffee as willingly as real coffee, it is sufficient to add a few coffee beans to the acorns before roasting them.

In the treatment of phthisis, fever is the symptom which principally demands attention whenever it persists at all severely, in spite of the remedies directed against the main

disease.   Digitalis and quinia have a well-merited reputation
as means of arresting the abnormal calorification, and reducing the animal heat in spite of the continuation of the
disease.

```
341.  R.   Pulveris digitalis,              gr. x.
           Pulveris ipecacuanhæ.
           Pulveris opii,                aa gr. v.
           Extracti helenii,             q. s.          M.
        For xx pills.
     One three times a day.
```

Add quiniæ sulphatis ℈j. to the above prescription when
the type assumed by the fever becomes periodical, the evening
exacerbations severe, and the chills, by which they are usher-
ed in, pronounced.

Our author is so much in the habit of using this (known as
Heim's) pill, with or without quinine, in consumption, when
the fever proves refractory to other remedies, that it has become a very common prescription at his clinic.   The exhibition is suspended whenever a distinct reduction of the temperature and of the frequency of the pulse becomes apparent,
and is resumed so soon as the effect subsides.   Patients pretty
soon learn to judge for themselves when it is time to stop the
pills, and when to resume them.*

DR. DOUGLAS POWELL, HOSPITAL FOR CONSUMPTIVES.

BROMPTON.

```
342.  R.   Potassæ chloratis,             ℈ij-iij.
           Morphiæ muriatis,              grs. jss-ij.
           Glycerinæ,                     f.℥ss.
           Syrupi,                     ad f.℥iv.          M.
```
A teaspoonful to be swallowed slowly in the rawness of the
   tongue, and painful deglutition in advanced phthisis.

It acts locally on the parts affected, relieving at the same
time the cough.   Of course, in the latest stages of the disease

*Text-Book of Practical Medicine, Am. Ed., 1869, vol. i, p. 244.

this will but render the remaining hours of life more comfortable; but there are some cases where this condition of the tongue and throat will come on earlier, and by rendering the taking of nutrients or stimulants almost impossible, cause death from exhaustion. In such instances, the above combination relieves pain, cleanses the tongue, and enables the patient to take nourishment and remedies which greatly prolong life.

JOHN C. THOROWGOOD, M. D., ETC., LONDON.

343. R. Sodæ hypophosphitis,   gr. v.
          Glycerinæ,
          Aquæ,   aa f.ʒss.
      For one dose, ter die

In addition to Dr. Thorowgood, Drs. C. J. B. WILLIAMS and C. J. WILLIAMS, (the latter, one of the physicians at the Brompton Hospital), speak of the value in phthisis of the *hypophosphites*, which at first so highly lauded by Dr. CHURCHILL of Paris, have of late fallen somewhat into disrepute.

THOMAS HAWKES TANNER, M. D., F. L. S., ETC.

344. R. Ferri iodidi,   gr. vj–xviij.
          Glycerinæ,   f.ʒij.
          Infusi calumbæ,   q. s. ad f.ʒvj.   M.
      Two tablespoonfuls three times a day.

In strumous ulcers, etc., where the stomach will not tolerate cod-liver oil, the above is useful.

345. R. Ammonii iodidi,   gr. j–vj.
          Infusi cinchonæ flavæ,   f.ʒss.   M.
      For one dose, to be taken twice or thrice daily, before food.
      Very valuable in strumous enlargement of the absorbent
      glands. The dose is to be graduated according to the
      patient's age. At the time the medicine is given intern-
      ally, the following ointment should be rubbed into the
      swelling night and morning:

346. R. Ammonii iodidi,   ʒj
          Adipis,   ʒj.   M.

In cases of phthisis, where the stomach will not tolerate any form of cod-liver oil, resort may be had to *cod-liver oil embrocations* :

347.   R.   Olei morrhuæ,   f ℥iijss.
            Spiritûs ammoniæ aromat., f ʒj.
            Tincturæ opii,   f.℥ss.
            Olei lavandulæ,   ℳxxx.   M.

One-half to be rubbed over the chest and abdomen, night and morning. Or,

348.   R.   Olei morrhuæ,   f.ʒj.
            Olei cajuputi,   f.ʒj.   M.

To be rubbed over the chest at bed-time, and applied by means of lint well saturated with it. The cajuput oil well disguises the smell of the embrocation.

349.   R.   Ferri ammonio-sulphatis,   ʒss–ʒj.
            Aquæ destillatæ,   f.℥vj.   M.

Two tablespoonfuls every six or eight hours, in cases where, on account of hæmoptysis, an astringent preparation of iron is indicated.

350.   R.   Liquoris potassæ,   f ʒiij.
            Tinct. cinchonæ compositæ, f.ʒvj.
            Decoc. cinchon. flav., q.s. ad f.℥vj.   M.

Two tablespoonfuls twice or thrice daily.

Often beneficial in the early periods of the disease. But it is a less favorite remedy with our author than

351.   R.   Spiritûs ammoniæ aromat.,
            Spiritûs chloroformi,   aa f.ʒvij.
            Morphiæ muriatis,   gr.j.
            Extracti cinchonæ fluidi,   f.℥ss.
            Tincturæ cinchonæ,   q.s. ad f.℥iij.   M.

One teaspoonful in a wineglass of port wine, three times a day.

In certain cases of phthisis this mixture is very useful, especially in conjunction with cod-liver oil and a liberal diet.

If the *night sweats* weaken and annoy the patient, they may be treated with

352. ℞. Acidi gallici, ℈ij.
Extracti cannabis indicæ, gr. v.
Confectionis rosæ, gr. x. M.
For x pills.
One to be taken every night at bed-time. Or,

353. ℞. Zinci oxidi, gr. xij.
Extracti conii,
vel,
Extracti hyoscyami, gr. xviij. M.
For vj. pills.
One to be taken every night at bed-time.

For the relief of night sweats in phthisis and other exhausting diseases there are few remedies more serviceable than the foregoing.

___

PNEUMONIA.

WM. AITKEN, M. D., EDIN.

354. ℞. Antimonii et potass. tart., gr. ½-j.
Hydrargyri chloridi mitis, gr.j. M.
For one pill. To be given every five or six hours, according to the severity of the disease.

This combination is believed to have saved a much larger number of cases than antimony alone. It is to be adopted in some cases. The bowels should be well cleaned out before resorting to it. So soon as the gums are touched, the prescription should be discontinued.*

M. BOUCHUT.

355. ℞. Veratriæ,
Pulveris opii, aa gr.jss.
Pulveris ipecacuanhæ, gr.iij.
Syrupi, q. s. M.
For xx pills.
From one to five to be taken during the day.†

___

* Science and Practice of Medicine, Am. Ed.,vol. ii. p. 756.
†O. Reveil, Formulaire Raisonné des Médicaments Nouveaux, p. 251

J. M. DA COSTA, M. D., PHILADELPHIA.

356. ℞. Potassii iodidi,   Əiv.
          Tinct. cinchoniæ compositæ, f.℥iv.   M.

A dessertspoonful ter die, in sub-acute pneumonia with pleurisy. Also,

357. ℞. Emplastri cantharidis,   4x5.

To be followed by a poultice and dressed with basilicon ointment.

THOMAS HILLIER, M. D., LOND., F. R. C. P., ETC.

LOBULAR PNEUMONIA OF CHILDREN.

Usually the best treatment is to keep the patient in bed in a room about 60°, well ventilated, without a draft, milk diet during the height of the fever and when the temperature falls some good beef-tea, and a simple saline mixture, such as

358. ℞. Potassæ citratis,   Əj.
          Syrupi aurantii,   f ℥ij.
          Aquæ,   q. s. ad f.℥ij.   M.
    Two teaspoonfuls pro re nata.

The tendency of the disease in children is to recovery. The great point is to do nothing which will interfere with rapid convalescence. Antimony is seldom desirable or necessary; if given at all it should be confined to those cases in which the pulse is full and strong, the temperature very high, and the skin and mucous membranes very dry and injected, and it should be given only for a short time at an early stage of the disease. Counter-irritation is not much to be relied upon. When there is severe pain in the side, a mustard plaster is of service. Blisters are seldom or never to be recommended, certainly not in the acute stage. If resolution comes on very slowly, and there is persistent pleuritic pain, an occasional flying blister will be of service. Calomel is not to be recommended, except as an occasional aperient. If the pneumonia is complicated with bronchitis, and the bronchia contain much mucus, a stimulant expectorant is indicated, such as

359. ℞. Ammoniæ carbonatis, gr. viij$xij.
Tincturæ scillæ, ℳ.xx.
Syrupi, f.ℨij.
Decocti senegæ, q. s. ad. f.ℨij. M.
Two tea-spoonsful for a child three years old.

During convalescence the use of iron, in a mild form, is of service ; as

360. ℞. Ferri et quiniæ citratis, Əj.
Syrupi limonis, f.ℨij.
Aquæ, q. s. ad. f.ℨij. M.
Two teaspoonfuls ter die.*

DR. FELIX VON NIEMEYER, PROFESSOR UNIVERSITY OF TUBINGEN.

361. ℞. Quiniæ sulphatis, Əj.
For x pills. One every two hours.

According to experiments of our author in the administration of quinine in this disease, it is called for when there is great danger, arising chiefly or entirely from excessive elevation of the temperature of the body. It may then be given as above, or in two or three ten-grain doses within a few hours.

DR. JOHN POPHAM, PHYSICIAN TO THE CORK NORTH INFIRMARY.

362. ℞. Potassæ bicarbonatis, ℨj-vj.
Misturæ acaciæ, f.ℨiij. M.
A dessertspoonful in water, four, six or eight times in the twenty-four hours.

The evidence of the good effects of this alkaline treatment appears on the second or third day. It acts as a sedative by allaying the cough and abating the state of congestion on which it depends. A blister applied for four or six hours, but not for suppuration, is a valuable auxilliary. Suppuration from blistering is exhaustive and prejudicial.†

*Diseases of Children, Am. ed., 1868, p. 48.
†*British Medical Journal* for Dec. 28th, 1839.

EUSTACE SMITH, M. D., LONDON, M. R. C. P., ETC.

363. R. Liquoris ammoniæ, acetatis, f ʒiv.
      Potassæ nitratis,   ʒj.
      Potassæ bicarbonatis,   ʒiss.
      Spiritûs ætheris nitrosi,   f.ʒiss.
      Aquæ carui,   q. s. ad. f.ʒvj.   M.

A tablespoonful every third hour for a child six or seven
years old. In cases of "pulmonary phthisis." At the
same time the child should be kept covered with hot linseed meal poultices, frequently renewed; and the child
confined to his nursery or bed-room.

After the cough has become loosened and the oppression of
the chest has subsided, expectorants should be given with an
alkali:

364. R. Spiritûs ammoniæ aromatici,
      Spiritûs ætheris nitrosi,
      Vini ipecacuanhæ,   aa. f ʒj.
      Potastæ bicarbonatis,   ʒj.
      Infusi calumbæ,   q. s. ad. f.ʒvj.   M.

A tablespoonful every sixth hour.

Afterwards when the secretion is free, easily brought up,
and the fever has disappeared, an astringent is useful, combined with expectorants and a little opium:

365. R. Liquoris ferri pernitratis,
      Acidi nitrici diluti,   aa. f ʒj.
      Tincturæ opii camphoratæ,   f.ʒij.
      Oxymellis scillæ,   f.ʒj.
      Infusi calumbæ,   q. s. ad. f.ʒvj.   M.

A tablespoonful ter die.

When unabsorbed pneumonic deposits continue, alkalies are
extremely useful. The inhalation of sprays of weak solutions
of bicarbonate, nitrate or chlorate of potash promotes expectoration; thus, for

INHALATION.

366. R. Potassæ bicarbontais,   gr.x.
      Aquæ,   f.ʒj.   M.

To be inhaled, by atomization, twice a day.

At a later stage an astringent spray may be used, such as

368. R. Acidi tannici, gr. iij.
Aquæ, f.ʒj. M.
For atomization, twice a day.

In chronic pneumonic consolidation irritants are sometimes useful :

368. R. Olei tiglii, f.ʒj.
Linimenti saponis, f.ʒj. M.
To be rubbed into a limited spot on the chest twice a day till pustulation, and then once a day for a week.

So long as there is much heat of the skin counter-irritants should not be employed.*

ACUTE PLEURISY.

PROF. ROBERTS BARTHOLOW, CINCINNATI, OHIO.

HYPODERMIC INJECTION.

369. R. Morphiæ sulphatis, gr. xvj.
Aquæ destilatæ, f.ʒj. M.
Dissolve and filter.
Dose—five to ten minims.

Nothing can be more satisfactory than the treatment of pleurisy *in its early stage* by the hypodermic injection of morphia. It relieves at once the pain, and arrests or diminishes the morbid process.†

DR. J. M. DA COSTA, PHILADELPHIA.

370. R. Potassæ acetatis, gr. xv.
Spts ætheris nitrosi, f.ʒss.
Vini ipecacuanhæ, gtt. iij
Syrupi tolutanus, f.ʒss. M.
For one dose, four times a day. Useful in sub-acute pleurisy.

---

*Wasting Diseases of Infants and Children, Am. Ed., p. 171, 1870.
†Manual of Hypodermic Medication. Lippincott, 1869, pp. 36, 61.

8

371.   ℞.   Tincturæ veratri viridi,   ♏xxiv.
           Potassæ acetatis,   ℥ss.
           Morphiæ acetatis,   gr.ss.
           Liquoris potassæ citratis,   f.℥ijss.
           Syrupi tolutanus,   f.℥ss.   M.

A dessertspoonful every three hours in dry pleurisy.

Locally, apply two or three times a day, turpentine stupes.

CHRONIC PLEURISY.

WM. AITKEN, M. D., EDINBURGH.

372.   ℞.   Pulveris digitalis,
           Pulveris scillæ,
           Pilulæ hydrargyri,   aa   gr iss.   M.

For one pill, two or three times a day, as a diuretic in
   chronic pleuritic effusion. Also

373.   ℞.   Hydrargyri chlo. corrosivi, gr. iv.
           Tinct. iodinii compositæ,   f.ʒiv-vj.
           Glycerinæ,   f.℥iij.
           Aquæ, distillatæ,   f.℥ivss.   M.

For a lotion to be applied over the chest by spongeopiline,
   or by lint covered with oiled silk.

One or other of the following ointments may also be rubbed
in upon the skin, over the side of the chest, namely:

374.   ℞.   Hydrargyri chlor. cor.,   gr. iv-v.
           Unguenti iodinii compos.,   ʒiv-ʒvj.
           Adipis,   ʒiv-℥j.   M.

Or

375.   ℞.   Hydrargyri chlor. cor.,   gr. iv-v.
           Potassii iodidi,   ʒij.
           Aquæ destillatæ,   q. s. to make solution.
           Adipis,   ℥j.   M.
           Ft. ung.*

---

*Science and Practice of Medicine, vol. 2, p. 682.

J. M. DA COSTA, M. D., PHILADELPHIA.

376. R. Potassii iodidi,   gr. viij.
          Ext. pruni virginianæ fl.,   f.ʒj.
          Spts. juniperis compositi,   f.ʒiij.   M.
For one dose, ter die.

377. R. Potassæ acetatis,   ʒj.
          Tincturæ digitalis,   f.ʒij.
          Extracti cinchonæ fluidi,   f.ʒj.
          Aquæ,   f.ʒij.   M.
A teaspoonful to die in pleuritic effusions.

378. R. Potassii iodidi,   ʒij.
          Tincturæ scillæ,   f.ʒvj.
          Tincturæ opii camphoratæ,   f.ʒjss.
          Misturæ acaciæ,   f.ʒvj.   M.

A teaspoonful four times a day, in chronic pleurisy with
  consolidation of the lung. Also a good nourishing diet ;
  either whisky or gin, half an ounce three times a day ;
  and counter-irritation by means of tincture of iodine.

Afterward, when effusion has begun to disappear, the following tonic-diuretic may be ordered :

379. R. Tincturæ ferri chloridi,   f.ʒj.
          Acidi acetici,   f.ʒj.   M.

And add :

          Liquoris ammoniæ acetatis,   f.ʒv.
          Syrupi aurantii corticis,   f.ʒij.   M.
A dessertspoonful increased to a tablespoonful, ter die.

380. R. Potassii iodidi,   ƺiv.
          Potassæ acetatis,   ʒss.
          Elixir. cinchonæ,   f.ʒiij.
          Curaçao,   f.ʒj.   M.

A dessertspoonful ter die in pleuritic effusion, with rough
  ening above effusion. Also a blister and an occasional
  catharic.

THOMAS HAWKES TANNER, M. D., F. L. S., ETC.

381.　℞.　Pilulæ hydrargyri,　　　gr. iij.
　　　　　Pulveris digitalis,　　　　gr. ss.
　　　　　Pulveris scillæ,　　　　　gr. jss.　　　　M.
　　　For one pill.

To be taken as an alterative and diuretic, two or three times
a day. (The dose in this pill differs somewhat from the
same combination given by Dr. Aitken. F. 372.)

Very often, however, mercury in any shape does harm.
Then, the compound tincture of iodine, the iodide of iron, or
cod-liver oil, are much more likely to be useful.

The patient ought to be kept on a moderate diet, free from
stimulants. A series of flying blisters may be applied. Purgatives as well as diuretics should be administered.*

DR. CHARLES WEST, LONDON.

382.　℞.　Potassii iodidi,　　　　gr. xij.
　　　　　Potassæ nitratis,　　　　gr. xxx.
　　　　　Spiritûs ætheris nitrosi,　f.ʒj.
　　　　　Tincturæ scillæ,　　　　♏xxx.
　　　　　Tincturæ digitalis,　　　♏xxiv.
　　　　　Syrupi aurantii,　　　　f.ʒss.
　　　　　Aquæ,　　　　　　　　ad f ʒiv.　　　M.
　　Sig.—Tablespoonful every four hours, for a child six years
　　old.

Employed in the treatment of pleuritic effusion, and continued steadily for several days. Its action may be seconded
by a small dose of mercury given once or twice a day, as one
grain of calomel, or three of gray powder. The mercury
may be discontinued at the end of a week, but the iodide of
potassium may be persevered with for two or three weeks.

*Practice of Medicine, p. 359, Am. Ed.

III. DISEASES OF THE CIRCULATORY SYSTEM.

FUNCTIONAL PALPITATION OF THE HEART.

J. M. DA COSTA, M. D., PHILADELPHIA.

Inquiry should always be made as to the cause, for the first step in the treatment is its removal. The cause may be found to be the drinking of coffee, chewing of tobacco, smoking, alcoholic drinks, masturbation, etc.

In all cases of functional disorder of the heart, attended with palpitation, digitalis is very serviceable, more so than aconite. If there be masturbation as the exciting cause, the following is a useful combination:

383. R. Potassii bromidi, ℨvss.
     Tincturæ digitalis, f.ℨijss.
     Infusi cascarillæ, f.ℨiv. M.
A desertspoonful two or three times a day

CARDIAC HYPERTROPHY.

J. M. DA COSTA, M. D., PHILADELPHIA.

In the treatment of hypertrophy of the heart, as much rest as possible should be insisted upon. The patient must be directed to lie down for several hours each day. The sinking of the pulse, which naturally occurs in the recumbent position, makes this posture as potent a cardiac sedative as we possess. All stimulants to the action of the heart should be removed. This includes the removal of any dyspeptic symptoms which may be present, and of any other disordered function which can react upon the heart.

There are only two drugs in which our author has any confidence; one is aconite, or its active principle, aconitia, and the other, veratria. These are the only medicines which directly and positively control the element of muscular power of the heart. Digitalis does not compare as a pure sedative with either aconite or veratrum viride. Gelseminum is useless; it has a false reputation. It is time lost to employ it. Hydrocyanic acid is often a useful and pleasant adjunct, when there is gastric disturbance, alone it is of no value.

These principles apply both in the treatment of simple hypertrophy, and in that complicated with valvular disease. A certain amount of hypertrophy with valvular disease is beneficial, and judgment must be exercised in order to determine when to interfere, and when not.

384. R. Tinct. veratri viridis,   f.ʒjss.
         "   aconiti radicis,   f.ʒss.
         "   zingiberis,   f.ʒvss.   M.
Fifteen drops ter die, two hours after meals in water.

The addition of the tincture of ginger causes the veratrum viride to be better borne by the stomach. Our author also frequently orders

385. R. Tinct. aconiti radicis,   gtt.j.
Ter die *for many months*, its effects being watched.

In a large number of cases this remedy thus employed prevents the further growth of the heart, and in some it lessens the already existing bulk of the organ.

DR. WALSHE.

386. R. Extracti aconici alcoholiti, gr. ijss.
For xx pills.
One as a dose. In repeating the doses, the effects must be watched.

Our author prefers this medicine to all other cardiac sedatives in this affection. He gives the preference to this preparation of the drug over the tincture. It relieves the painful sensations and disquietude about the heart.

CARDIAC DILATATION.

J. M. DA COSTA, M. D., PHILADELPHIA.

387. R. Pulveris digitalis, gr. v.
Extracti belladonnæ, gr. j.
Ferri redacti, Ɔij. M.
For xx pills.
One ter die.

388. R. Emplastri belladonnæ, 4x4.
To be worn over the cardiac region.

Belladonna is one of the best agents that can be employed to overcome irregularity of the action of the heart, and to relieve pain. Digitalis is also useful for the same purpose, especially where the action of the heart is feeble; it is the only sedative which will reduce the frequency and not the force of the heart. It may be combined as follows:

389. R. Ferri lactatis, ʒss.
Pulveris digitalis, gr. v. M.
For xx pills. One ter die.

390. R. Tincturæ digitalis, f.ʒss.
Ten drops ter die, in cases of dropsy caused by cardiac dilatation. Also,

391. R. Pulveris jalapæ compositus, gr. x.
For one dose at night.

In such cases, a tablespoonful of lemon juice, three times a day, acts as a diuretic and stomachic. Baths, also, are advantageous. Dry cups applied to the chest relieve the pulmon-

ary congestion. It is more important to start the secretions
and relieve internal congestions than to give tonics and iron
—which find their appropriate place in the after treatment.

DR. FELIX VON NIEMEYER, PROF. UNIVERSITY OF TUBINGEN.

Our author has convinced himself, by a great number of
observations, that digitalis is a very efficient means of temporarily strengthening the heart's contractile power, and of thus
allaying cyanosis and dropsy. In dilatation of the heart,
digitalis, combined with an exclusively milk diet, is an invaluable remedy. Dr. von N. has repeatedly succeeded in
obtaining complete removal of dropsical effusions of great
magnitude and producing considerable temporary relief by
this mode of treatment.

Iron, which fortunately no longer has the reputation of
being " heating," should always be prescribed when the patient
shows any signs of anæmia or hydræmia.*

DR. WALSHE.

When dropsy appears in cases of dilatation of the heart,
the diuretics which yield most relief are the acetate, nitrate,
iodide, and bitartrate of potassa, nitrous ether, compound
tincture of iodine, the infusion and spirits of juniper, and gin.
Hydragogue cathartics, elaterium, gamboge, citrate of potassa
and the compound jalap powder, also aid in subduing the
dropsical effusion. The following formula is a useful one for
the administration of elaterium ;

392.   R.   Extracti elateri,   gr.1-6½.
            Extracti creasotonis,   gr.ij.
            Extracti hyoscyami,   gr.ij.   M.
       For one pill.

---

*Text Book of Practical Medicine, Am. ed. vol. i., p. 325. 1869.

CARDIAC REGION.

ROBERTS BARTHOLOW, A.M., M. D., ETC., CINCINNATI, O.

In that form of *angina pectoris*, which is essentially a neuralgic affection of the cardiac nerves, our author has had very satisfactory results from the hypodermic administration of morphia. (F. 369.)

The violent and irregular actions of the heart occurring in hysterical subjects are immediately relieved by the use of the hypodermic syringe. Morphia alone is used. In cases of dyspnœa, dependent upon dilated right cavities, pulmonary œdema and mitral disease, advantage is derived from the following

HYPODERMIC INJECTION.

393. ℞. Morphiæ sulphatis,   gr. xvj.
          Atropiæ sulphatis,   gr. j.
          Aquæ destillatæ,   f.℥j.   M.
      Filter.

Dose—five minims (equal to one-sixth of a grain of morphia, and one ninety-sixth of a grain of atropia).

Our author has not observed any good effects from hypodermic medication in hypertrophy and semi-lunar disease of the heart. Violent palpitations, produced by emotion or reflex irritation, in cases of organic disease of the heart, may be palliated by the subcutaneous use of morphia; but his own observations are unfavorable to the employment of hypodermic injections in narrowing and obstruction of the aortic orifice.*

INTERNAL ANEURISM.

J. M. DA COSTA, M. D., PHILADELPHIA.

There are only two remedies in which our author has any faith in the radical treatment of internal aneurism. The first

---

*Manual of Hypodermic Medication, p. 62.

is *iodide of potassium*. It should be used boldly. The following recipe was given continuously for ten months, with the most marked beneficial results, in a case of thoracic aneurism :

394. R. Potassii iodidi,   gr. xv.
           Syrupi tolutanus,
           Aquæ,   aa f.ʒj   M.
      For one dose, ter die.

This remedy does no good excepting early in the disease. The second remedy referred to, is *ergot.* It is not yet known definitely how much good it really does. Some very excellent results have been obtained by Langenbeck. It may be given internally or by hypodermic injections.

In a disease so dangerous, so almost necessarily fatal, the importance of a knowledge of any remedy which seems to exert an influence is apparent. As both the iodide of potassium and ergot can be tried without injury to the patient, it is the duty of every practitioner, in cases of internal aneurism, (in which, of course, surgical treatment is out of the question,) to try one or the other of these drugs.

The following formula may be used for the hypodermic injection of ergotin :

395. R. Ergotini,   gr. ij.
           Spiritûs vini rectificatæ,
           Glycerinæ,   aa f.ʒss.   M.
      Five minims (equal to gr. 1-6 of ergotin) for a dose. This
      is the formula of Eulenberg.

Prof. Langenbeck employs the aqueous extract of ergot or *Bonjean's ergotin.* It is usually administered hypodermically in the dose of gr. ¼. In a case reported by Langenbeck, thirty grains of this preparation were injected in forty days with great benefit. The subclavian aneurism diminished in size, and the other symptoms improved. A case of radial aneurism was speedily cured by the hypodermic injection of this preparation over the tumor.

Prof. BARTHOLOW states* that the following formula may be used:

396. R. Ext. ergotæ fluidi (U. S. P.), f ʒij.
Carefully filter and inject in doses of ten minims.

IV. DISEASES OF THE DIGESTIVE APPARATUS.

THE MOUTH.

WILLIAM AITKEN, M. D., EDINBURGH.

397. R. Liquoris ferri pernitratis, gtt.x.
Syrupi aurantii, f.ʒss.
Aquæ, f.ʒvss. M.

A fourth part to be given to a child three or four years of age four times a day, in *aphthous stomatitis.*

In cases where parasitic vegetable productions abound, the application of the following solution removes the lesions in twenty-four hours:

398. R. Sodæ sulphitis, ʒj.
Aquæ, f.ʒj. M.

The acid secretions of the mouth decompose the salt and set free the sulphurous acid, which destroys the parasite.†

399. R. Acidi carbolici fluidi, f.ʒss.
Aquæ bullientis. O.viij. M.

Allow the solution to become warm or tepid, and syringe the mouth frequently with it in *cancrum oris.*

*Manual of Hypodermic Medication, p. 127.
†Science and Practice of Medicine, Am. Ed., vol. ii., p. 817.

PROF. S. D. GROSS, PHILADELPHIA.

400. ℞. Liquoris plumbi subacetatis, f.ʒj.
Aquæ,   f.ʒviij.   M.
To be used as a mouth wash every hour or two in cases of *mercurial stomatitis.*

The only objection to this lotion is that it discolors the teeth, which effect, however, quickly disappears. At the same time, internally, the chlorate of potassa should be administered. Fifteen to thirty grains are to be taken ter die in mucilage or lemonade.

PROF. J. LEWIS SMITH, NEW YORK.

401. ℞. Potassæ chloratis,   ʒj.
Mellis,   f.ʒss.
Aquæ,   f.ʒij.   M.
One teaspoonful every two or three hours, in *ulcerous stomatitis.*

It often acts like a specific for this as well as other forms of stomatitis. It should be allowed to run over the affected part, as it is believed to have a local action.*

402. ℞. Sodæ biboratis,   ʒj.
Glycerinæ,   f.ʒj.   M.
This wash is to be applied by a camel-hair pencil, or with a soft cloth upon the finger or a stick, four or five times daily in *thrush.*

There is an objection to using any application for the removal of thrush which contains either sugar or honey, since either substance remaining in the mouth would rather promote the growth of the parasite.

In the intervals, between the applications of borax, if the buccal surface be hot, dry and tender, mucilaginous washes, as the mucilage of acacia or mallow should be employed. If the disease continue, the mouth should be occasionally washed with

---

*Treatise on the Diseases of Infancy and Childhood. 1869, p. 288

103. R. Zinci sulphatis, gr. ij-iv.
     Aquæ rosæ, f. ℥ij, M.

404. R. Cupri sulphatis, ℥ij.
     Pulveris cinchonæ, ℥ss.
     Aquæ, f ℥iv. M.

To be applied very carefully twice a day, to the full extent of the ulcerations and excoriations in *gangrene of the mouth*.

This local treatment (recommended by Evanson and Maunsell), our author believes to be preferable to that advised by any other writer. He has seen it so successful that he should employ it in all ordinary cases from the first visit.

The addition of cinchona is useful in this formula by retaining the sulphate of copper longer in contact with the edges of the gums:

In some cases the following is useful:

405. R. Zinci sulphatis, ℥j.
     Aquæ, f.℥j. M.

The above treatment is preferable, provided it is equally effectual in arresting the gangrene, to the application of the strong escharotics recommended by many.

ALFRED VOGEL, M. D., PROF. IN THE UNIVERSITY OF DORPAT, RUSSIA.

406. R. Sodæ biboratis, Əj.
     Aquæ, f.℥j. M.

To be used to cleanse the mouth every hour in cases of *stomatitis* in infants.

This feebly alkaline solution combats the tendency of the profusely secreted saliva to rapidly become sour. The chest is to be protected against getting wet by a piece of oil silk, which is secured under the jacket, and the infant is only to be allowed to drink cow's milk with water. The painful ulcers may be relieved for many hours, and even permanently, by cauterizing them with the solid nitrate of silver.

In idiopathic stomatitis spontaneous recovery takes place in eight, or at the longest, fourteen days. Symptomatic stomatitis in febrile disease, does not usually call for any particular interference.*

407. ℞. Potassæ chloratis, Əj.
Syrupi, f ℥ij.
Aquæ, q. s. ad f.℥iv. M.

The whole amount to be administered in the course of the day to a child, one year of age, in *putrid sore mouth.*

At the end of this time the smell, in all cases and in every degree of the affection, is *completely abolished.* The remedy should be continued three or four days, or the disease will return. Our author has never found it necessary to employ this remedy longer than four days, nor has he ever noticed any bad effects, such as loss of appetite, diarrhœa, etc.

For a child, over one year and under two, the chlorate of potassa, in the above formula, should be increased to ʒss.; for a child under three, to Əij. Children who have attained the fourth year tolerate very well ʒj. pro die.†

THE FAUCES.

J. M. DA COSTA, M. D., PHILADELPHIA.

408. ℞. Cupri sulphatis, ʒj.
Aquæ, f.℥j. M.

Apply with a brush three times a week in cases of *follicular pharyngitis.*

W. KEMPSTER, M. D., UTICA, NEW YORK.

GARGLE.

409. ℞. Acidi carbolici, gr.viij.
Aquæ, f.℥iv. M.

Use as a gargle in cases of *common sore throat.* It has the advantage over the ordinary potassa gargles of relieving the bad taste and foul breath.

---

*Practical Treatise on the Diseases of Children.. Am. ed., 1870, p. 88.
†Ibid, p. 95

PROFESSOR JOSEPH PANCOAST, PHILADELPHIA.

410. R. Cinchonæ rubri, ℥ss.
Aquæ, bullientis, Oss. M.
Strain and add:
Tincturæ myrrhæ,
Tincturæ krameriæ,
Mellis despumatæ, aa f.℥j.
Acidi muriatici diluti, gtt. xv. M.

Use as a gargle in cases of *chronic sore throat.*

411. R. Myrrhæ, ℥ij.
Sodæ biboratis, ℥j.
Mellis, f.℥j.
Aquæ bullientis, Oss. M.

To be drawn through the nostrils into the mouth in cases of
irritation of the back part of the nostrils and of the velum
pendulum palati.

ENLARGED TONSILS.

A. RUPPANER, M. D., NEW YORK.

In cases of chronic enlargement of the tonsils, our author
uses the *London paste,* recommended by Dr. MORRLL MAC-
KENZIE. He gives the following directions for its preparation :

412. R. A quantity of equal parts of finely pulverized and
well mixed *caustic soda* and *unslacked lime* is kept on hand.
When an application is to be made to the tonsils, a little
of the powder is put into a small porcelain cup, a few
drops of absolute alcohol, which is kept near at hand,
are added; the two are carefully mixed with a glass rod;
when the paste is ready for use. Care must, however, be
taken that it be of the proper consistency. If too thin it
is apt to find its way to parts which ought not to be touched; if too thick or lumpy, the paste will not readily stick,
and little pieces might be swallowed. To apply the paste,
a glass rod of sufficient length ought to be used. One end
of it, which must be smooth and slightly funnel shaped, is
dipped into the paste, and a greater or lesser portion of
the surface touched, as occasion may require :

To apply the paste the patient should be placed in the
position for laryngoscopy. The tongue is then to be depressed

with the spatula, and the paste applied to the enlarged surface for two or three seconds. The action of the escharotic upon the tonsil is rapid. The mucous membrane almost instantly assumes a deep flesh color, and presently a dark blackish spot is seen streaked with blood. The following day the tonsil is covered with a whitish-yellow eschar.

The inconsiderable amount of suffering produced by this application is noticeable. Children hardly pay any attention to the pain, or make light of it. At the longest, the discomfort lasts only about two or three minutes. Subsequent applications are accompanied with less, if any pain at all.

The operation is again to be repeated in two or three days. The number of applications will depend upon the nature of the case.

Our author reports one hundred and twenty-three cases treated in this manner—the minimum number of applications of the paste, in any case, was six, the maximum, fourteen. This new escharotic does away with the necessity of resorting to the knife for the removal of enlarged tonsils.

DIPHTHERIA.

WM. AITKEN, M. D., EDINBURGH.

413. R.   Extracti nucis vomicæ,
            Ferri sulphatis,   aa gr. v-x.
            Pilulæ rhei compositæ,   ℈ij-iij   M.
For xx pills. One morning and evening in the paralysis following diphtheria.*

THOMAS HILLIER, M. D., LOND., F. R. C. P., ETC.

414. R.   Hydrarg. chlor. mitis,   gr. iij-vj.
            Pulv. ipecac. comp.,   gr. v.   M.
      For six powders.
One every two or three hours for a child.

* Science and Practice of Medicine, Am. Ed. vol. i, p. 524,

Calomel is now almost discarded in the treatment of diphtheria. Our author is not prepared to give it up. In some
of his worst cases, in which recovery occurred, this drug was
the remedy. It is not to be used indiscriminately in all cases.
It should be limited to children with moderate constitutions,
and to cases in which the exudation is firm and thick, or
causing laryngeal obstruction with sthenic symptoms. It is
to be continued until the bowels are relaxed with greenish
stools. At the same time, abundant fluid nourishment and
sometimes wine are to be given.

PROF. J. LEWIS SMITH, NEW YORK.

415. R. Sodæ bisulphitis, ℥j.-ij.
    Tincturæ aurantii, f.℥ij.
    Aquæ, f.℥ x. M.

One teaspoonful every two hours. Sometimes, in place of
water, a bitter infusion like that of quassia, has been
employed.

The sulphites have not been employed sufficiently to determine their value in this disease. One author considers the
following mixture one of the very best for ordinary cases of
diphtheria :

416. R. Tincturæ ferri chloridi, f.℥j.
    Potassæ chloratis, ℥j.
    Syrupi simplicis, f.℥ ij. M.

One teaspoonful every two or three hours to a child of two
or three years.

No drinks should be allowed the patient for a few minutes
after each dose, in order that the full local effect may be
obtained.

In those of full habit and florid complexion, iron is not so
imperatively required. In such cases use the following :

417. R. Elixir cinchonæ, f.℥iv.

A teaspoonful to a tablespoonful for a dose, according to the
age. This is a useful and not unpleasant remedy.

9

The formulæ recommended in the topical treatment of the larynx in croup are proper for the pharynx and larynx in diphtheria (see F. 174, 175, 176). In those old enough the following is a useful

GARGLE.

418. R. Potassæ chloratis,   ℈iv.
          Aquæ,   f.℥iv.   M.
      Add to a teaspoonful of this:
          Tincturæ ferri chloridi,   f.℥j.
      And gargle with it every hour or two.

For the paralysis following diphtheria, the following formula (recommended by Professor METCALFE, of New York,) is useful:

419. R. Strychniæ,   gr. j.
          Acidi nitrici diluti,   f.℥j.
          Aquæ,   f.℥vij.   M.
      From three to five drops in a dessertspoonful of water are
          to be given three times daily to a child of three years.

The anæmic state which succeeds diphtheria is to be remedied by the administration of iron, for several weeks.*

THOS. HAWKES TANNER, M. D., F. L. S, LONDON.

420. R. Quinæ sulphatis,   gr. ij.
          Acidi muriatici diluti,   ♏x.
          Tinct. ferri chloridi,   ♏xv.
          Infusi calumbæ,   f.℥ss.   M.
      For one dose in water, ter die.

This preparation is also recommended by Dr. Aitken for the treatment of diphtheria.

_____

*Diseases of Infancy and Childhood, 1869, p. 457.

FUNCTIONAL INDIGESTION.

WILLIAM AITKEN, M. D., EDINBURGH.

421. R. Sodæ bicarbonatis,   gr. xv.
       Potassæ nitratis,   gr. iij.   M.

For one powder: to be taken two or three times a day, in those forms of indigestion marked by excessive acidity and heartburn. At the same time free excretion from the liver and bowels must be sustained by occasional small doses of blue pill or podophyllin, combined with extract of colocynth and of henbane, while exercise and diet are duly attended to.

422. R. Ammoniæ carbonatis,   gr. j.
       Extracti gentianæ,   gr. ij.   M.

For one pill, ter die in weakened digestion from over-fatigue.

423. R. Extracti nucis vomicæ,
       Ferri sulphatis,   aa gr. ss
       Ext. colocynthidis com., gr. iv.   M.

This combination taken early in the morning generally induces gentle action of the bowels.

In prescribing the mineral acids, our author calls attention to the following general rule, stated by Dr. BENCE JONES, namely, that the influence of sulphuric acid is astringent while that of muriatic acid promotes digestion, and of nitric acid secretion.

THOMAS KING CHAMBERS, M. D., CONSULTING PHYSICIAN AND LECTURER ON THE PRACTICE OF MEDICINE AT ST. MARY'S HOSPITAL, LONDON.

424. R. Acidi hydrocyanici diluti, ℳ iv.
       Infusi gentianæ,   f.ʒss.   M.

For one dose, ter die, in heartburn due to oversensitiveness.

425. R. Zinci oxidi,
       Pilulæ aloes et myrrhæ, aa. ʒjss.   M.

Divide into xx pills; one ter die, in the nervous trembling, indigestion of food and vomiting, arising from indulgence in spirit-drinking; between meals and in the forenoon.

J. M. DA COSTA, M. D.

426. R. Acidi nitro-muriatici, f.ʒij.
Vini pepsini, f.ʒiij.

A teaspoonful three times a day, before or after meals.

In functional indigestion owing to want of proper secretion of gastric juice. When there is constipation add also

427. R. Pulveris rhei, Ʒj.
Quiniæ sulphatis, gr.x. M.

Divide into x pills, one to be taken at night. If this be not sufficient to produce a laxative effect, take one night and morning. Meat diet almost exclusively, avoiding starchy substances.

PROF. JOSEPH PANCOAST, PHILADELPHIA.

428. R. Hydrargyri chloridi mitis, gr.¾.
Pulveris rhei, gr.j.
Pulveris zingiberis, gr.ss. M.

For one powder. To be taken by a child three years old, twice a week at night, to remove saburra, and as an alterative.

THOS. HAWKES TANNER, M. D., F. L. S., LOND., ETC.

429. R. Acidi nitro-muriatici diluti, f ʒij.
Acidi hydrocyanici diluti, ♏ xxv.
Tincturæ arnicæ, f.ʒj.
Tinct. gentianæ compositæ, f.ʒj.
Infusi sennæ, q. s. ad., f.ʒiij. M.

A tablespoonful two or three times daily, in dyspepsia with sluggish action of the liver.

The efficacy of this prescription may often be increased by giving with each dose the following pill:

430. R. Zinci sulphatis, gr.j-ij.
Extracti gentianæ, gr.iv. M.
431. R. Quiniæ sulphatis, gr.xij.
Pulveris ipecacuanhæ, gr.xij-xxiv.
Extracti gentianæ, gr.xxiv. M.

Divide into xij pills, and order one to be taken every day at dinner.

An excellent remedy in cases of slow digestion.

432. ℞. Ferri redacti,   gr.xxxvj-ʒj.
       Pepsinæ,   gr.xxxvj.
       Zinci phosphatis,   gr.xviij.
       Glycerinæ,   q. s.   M.

Divide into xxiv pills, silver them, and order two to be taken every day at dinner.

In anæmia, etc., with weakness of the digestive organs.

PROF. T. GAILLARD THOMAS, NEW YORK.

433. ℞. Magnesiæ sulphatis,   ʒij.
       Ferri sulphatis,   gr.xvi.
       Acidi sulphurici diluti, f.ʒj.
       Aquæ,   Oj.   M.

Two tablespoonfuls in a tumbler of ice water every morning upon rising, when a ferruginous tonic combined with a saline is indicated. Or,

434. ℞. Potassæ et sodæ tartratis, ʒij.
       Vini ferri amari,   f.ʒij.
       Acidi tartarici,   ʒiij.
       Aquæ,   f.ʒxij.   M.

Two tablespoonfuls in a tumbler of ice water before breakfast. Should this dose be not sufficient, two or three may be taken daily, for the result will prove tonic and reparative as well as cathartic.

435. ℞. One rennet, washed and
         chopped.
       Vini rubri,   Oj.

Macerate for twelve days, and then decant, filter, and add

℞. Acidi nitro-muriatici diluti, f.ʒij.
       Tincturæ nucis vomicæ,   f.ʒij.
       Bismuthi subnitratis,   ʒij.   M.

One tablespoonful in a quarter of a tumbler of water before each meal as a digestive tonic.

This prescription embraces the tonic properties of nux vomica and the peculiar restorative influence of bismuth, with a fluid which resembles the gastric juice. In many cases of habitual indigestion our author has obtained from it the best results.

436. ℞. Quiniæ sulphatis,   Əij.
        Ferri sulphatis,   Əj.
        Acidi sulphurici aromatici, gtt.x.
        Muc. acaciæ,   q. s.   M.
    Ft. massa. in pil. xx div.

One to be taken three times a day before each meal.

HABITUAL CONSTIPATION.

S. B. BIRCH, M. D., M. R. C. P., LONDON.

437. ℞. Extracti rhei alcoholici,   ʒss.
        Extracti taraxaci,   gr. xxiv.
        Quiniæ sulphatis,   gr. ij.   M.
    Ft. mass., in pilulas xij div.

One should be taken either on rising in the morning or at dinner time, or even at both periods when the constipation is very obstinate. This is a very gentle stomachic and tonic evacuant, particularly useful for the delicate. In addition, when there is torpor of the liver, deficiency or perversion of the biliary secretion, the patient should be ordered:—

438. ℞. Hydrargyri cum creta,   gr. ½-j.
        Sacchari albi,   gr. v.   M.
    For one powder.

A sufficient dose, when given alone at bed time, for two or three successive nights, or in very sensitive persons every second or third night. But the hyd. cum creta is often prepared imperfectly, and then causes considerable annoyance and dissatisfaction to the practitioner. It is better to prescribe it in the form given above, than in the form of a pill, which sometimes passes through the bowels unchanged.

In order to excite manifestation of contractile force on the

part of a torpid intestine, and for the conversion of irregular
and imperfect peristaltic action into a uniform and effective
power of expulsion, the following little operation executed
by the patient upon himself, when properly performed, is valuable:

Place the tips of the fingers of the right hand exactly
over the cœcal region with *very slight* pressure; carry them
upward along the ascending colon to the right hypochondrium;
continue the movement *without any intermission* over the region of the transverse colon to the angle of junction with the
descending colon; stop not, but proceed downward gently and
steadily to the iliac region of that side; instead of the previously very gentle pressure, the finger must now be pressed
firmly and deeply (without pain) into the pelvic cavity, and
there retained for about fifteen seconds; then remove the
hand altogether, rest a few seconds and repeat the procedure.
This may be continued for the period of from a few minutes
to a quarter of an hour or more. Some little care and tactile
dexterity is needed to do this properly, and where the hand is
dry or the cuticle thick and hard, it is advisable to slightly
moisten the ends of the fingers. When the right hand is
tired, the left can be used, and so alternately, but it is better
not to alternate them too rapidly. The patient may first try
and may succeed or not. If there be a failure, it must not
necessarily be given up. Invalids themselves will often fail,
almost invariably if their bowels be *extremely intractable.*
But now the aid of a friend for passive movements may be
invaluable. The medical adviser can give instructions re-
garding the precise anatomical relations of the parts involved,
and the method of performance. The proceeding should
usually extend over a period of from five or six minutes to
occasionally twenty-five minutes.

For occasional use in the treatment of constipation in old age, the following pill affords an excellent formula:

439. ℞. Ext. colocynth. compositi,   gr. v-viij.
          Extracti hyoscyami,   gr. ij.   M.

J. M. DA COSTA, M. D., PHILADELPHIA.

440. ℞. Podophyllin,   gr.1-20·
          Extracti belladonnæ,   gr.1-20.
          Capsici,   gr.¼.
          Pulveris rhei,   gr.j·   M.
For one pill. To be taken three times a day.

Belladonna is undoubtedly a stimulant to the muscular fibres of the intestine. It acts on them as it acts on the bladder; it stimulates to contraction. It also increases the action of purgatives, enabling the physician to get along with smaller quantities of purgative medicine. Podophyllin is useful in torpor of the upper portion of the bowel, to increase the secretion of the liver.

441. ℞. Tinct. gentianæ compositæ,   f.℥iij.
          Tincturæ rhei dulcis,   f.℥j.
          Tincturæ belladonnæ,   f.℥iss.   M.
Dessertspoonful ter die.

442. ℞. Extracti gentianæ,   gr.iij.
          Extracti nucis vomicæ,
          Podophyllin,   aa. gr.¼.
          Olei cajuputi,   gtt.j.   M.
For one pill. To be taken twice a day as a tonic for chronic constipation.

443. ℞. Extracti belladonnæ,   gr.1-16.
          Pulveris rhei,   gr.j·
          Pulveris zingiberis,   gr.ss.
For one pill. To be taken four times a day.   M.

Figs will sometimes act as purgatives when medicines fail.

PROF. ROBLEY DUNGLISON.

444. R. Magnesiæ sulphatis,   ℥j.
          Potassæ bitartratis,   ℥j.
          Ferri sulphatis,   gr. x.   M.

For one powder. Add to a quart of water and take a wine glass, on rising, every morning.

This recipe was frequently recommended by the late distinguished professor of physiology.

THOMAS HILLIER, M. D., LONDON, F. R. C. P., ETC.

In obstinate constipation occurring in children, very small doses of extract of belladonna, gr. 1-24 to gr. 1-32, will be found useful, and the abdomen may be rubbed with soap liniment and castor oil.

JOHN FORSYTH MEIGS, M. D.

445. R. Confectio. sennæ,   ℥j.
          Potassæ bitartratis,   ℥ij.
          Sulphuris præcipitati,
          Ferri subcarbonatis,   aa. ℥j.
          Mellis despumati,   q.s.   M.

Ft. electuary. Teaspoonful after meals.

PROF. METCALF, M. D., OF NEW YORK.

446. R. Extracti aloes (purif.),
          Extracti hyosciami,   aa. ℥j.
          Extracti nucis vomicæ,   gr. xij.
          Olei anisi,   gtt x.   M.

Divide into lx pills.

One to be taken after each meal, particularly for coconstipation in women.

THOMAS HAWKES TANNER, M. D., F. L. S., LONDON.

447. R. Zinci valerianatis,   gr. xij-xxiv.
          Extracti belladonnæ,   gr. iij-vj.
          Extracti gentianæ,   gr. xxiv.   M.

Ft. pil. xij and silver them.

One to be taken three times a day in nervous cases of habitual constipation, and in spasmodic contraction of the sphincter ani.

PROF. W. H. VAN BUREN, M. D., OF NEW YORK.

448. ℞. Extracti aloes, ℥ss.
Extracti nucis vomicæ, gr.vj.
Extracti hyoscyami, ℈j.
Pulveris ipecacuanhæ, gr.j. M.

Divide into xx pills. One to be taken at night.

This recipe, termed the "Pil. salutis," is of especial value in the constipation of females.

DIARRHŒA.

WILLIAM AITKIN, M. D., EDINBURGH.

449. ℞. Salicin, gr. v.

For one powder. To be taken every four or six hours.

In cases of diarrhœa with clean tongue, which will not yield to opiates, astringents, or stimulants, either singly or combined, and which probably depend on a want of tone in the intestine. In these cases the above recipe has often stopped a diarrhœa that appeared fast hurrying the patient to his grave.

One general rule may be acted on in the cure of diarrhœa, which is, that in the adult, whatever be the form of the diarrhœa, if the stools be dark at first, and then become light-colored, purgative medicines are no longer beneficial, and in no instance ought they to be continued longer than is sufficient to remove any irritative substance accumulated in the alimentary canal.

J. M. DA COSTA, M. D., PHILADELPHIA.

450. ℞. Bismuthi subnitratis, ℈i.
Acidi tannici,
Pulveris ipecacuanhæ compositæ, aa. gr. iij. M.

For one powder. To be taken three times a day in chronic dysenteric diarrhœa.

It is particularly in cases in which there exists persistent irritability of the bowels, influenced by the taking of much food which cannot be digested, and in which there are gastric symptoms in connection with the dysenteric affection, that the subnitrate of bismuth will be found very serviceable. In order that it shall produce an effect, it is necessary that it shall be administered in sufficiently large doses, not less than twenty grains. The dose may gradually be increased to a drachm.

THOMAS HILLIER, M. D., LONDON, F. R. C. P., ETC.

451. R. Acidi gallici,   gr.xij.
         Tincturæ cinnamomi,   f.ʒjss.
              "   opii,   ℳ viij.
         Aquæ carui,   q. s. ad. f.℥ij.

Dose—two teaspoonfuls for a child two years old, with chronic diarrhœa and irritable stomach.

452. R. Olei ricini,   f.ʒij.
         Pulveris acaciæ,   ʒj.
         Tincturæ opii,   ℳ viij.
         Syrupi,   f.ʒij.
         Aquæ carui,   q. s. ad. f.℥ij.   M.

Dose—A teaspoonful for a child six years old.

A useful oleagenous mixture in dysenteric diarrhœa.

W. KEMPSTER, M. D., UTICA, N. Y.

453. R. Acidi carbolici,   grj.
         Aquæ,   f.ʒj.   M.

This is the strength of the standard house solution in the State Lunatic Asylum at Utica, New York. Of this a dessertspoonful is given ter die, in cases of sluggishness of the bowels accompanied by offensive breath. Diarrhœa produced by eating unripe fruit, or other articles which promote fermentation, is relieved by combining a drachm or two of the solution with the usual remedies. When a fetid smell emanates from the cutaneous surface, order a warm bath, and then wash the surface with a solution—gr. v to f.ʒj.

JOHN FORSYTH MEIGS, M. D.

454. R. Pulveris opii,   gr.vj.
         Extracti nucis vomicæ,   gr.iij.
         Cupri sulphatis,   gr.j.   M.
    Ft. mass in pilulas xij, dividenda. One three times a day
       in chronic diarrhœa.

The value of this pill has been proved by army surgical
experience.

CHARLES MURCHISON, M. D., F. R. C. S., ETC.

455. R. Acidi tannici,   gr.x.
         Tincturæ opii,   ♏ v.
         Glycerinæ,   f.ʒss.
         Aquæ menthæ pip.,   ad. f.ʒss.   M.
    For one dose, in a tablespoonful of water every four hours.

After the diarrhœa is checked order:

456. R. Acidi nitro muriatici,   ♏x.
         Tincturæ opii,   ♏v.
         Syrupi,   f.ʒss.
         Aquæ,   ad. f.ʒss.   M.
    For one dose, in water, four times a day.

PROF. J. LEWIS SMITH, NEW YORK.

457. R. Tinctura opii camphorat⁓,
         Tinctura catechu,   aa. f.ʒij.
         Misturæ cretæ,   f.ʒj.   M.
    Dose—One teaspoonful every two to four hours to a child
       one year old.

Our author recommends the above as an old but efficient
prescription in the *simple diarrhœa of infants*.

458. R. Pulveris ipecacuanhæ,   gr. j.
         Pulveris rhei,   gr. ij.
         Sodæ bicarbonatis,   gr. iv–vij.   M.
    Divide in chart no. xij. One powder every four to six
       hours, to an infant one year old.

In the *diarrhœa of infants* due to indigestion and attended
by acidity. It improves digestion and corrects acidity, and
in this manner has a beneficial effect on the diarrhœa, in those
cases in which it is appropriate.

THOMAS HAWKES TANNER, M. D., F. L. S., ETC.

459. ℞. Cupri sulphatis,
Extracti opii,   aa. gr. ¼.
Extracti gentianæ,   gr. iij.   M.

For one pill, to be taken three times a day in obstinate diarrhœa.

460. ℞. Argenti nitratis,   gr. ss.
Extracti opii,   gr. ij.   M.

Make a pill to be taken night and morning.

In very obstinate diarrhœa where opium agrees with the system.

CHOLERA INFANTUM.

MEREDITH CLYMER, M. D., NEW YORK.

Our author gives the following indications for the treatment of *cholera infantum:*

The instant threatening symptoms, purging and vomiting, are to be stopped. Fermentation, and not chymification, is going on in the stomach and duodenum. Hence, small doses of the bisulphites of soda or potassa, with limed whey, will often act very happily, while the effect of poisonous drugs is always doubtful, and generally positively harmful. Mercury is, at best, negative. Opium and its preparations will be found valuable, if not contra-indicated by cerebral epiphenomena. The effects should be carefully watched. Flannel, wrung out of hot water, and on which laudanum is poured, applied to the spine, will be found useful in checking vomiting. The function of the skin which, in common with all the excreting organs, is inactive, must be excited. This may be done by gentle friction with woolen cloths, or a warm alkaline bath, in which the little patient should not remain longer than three minutes, being then quickly dried and wrapped in flannel. Food, of proper quality and quantity, should be given as soon as the stomach and bowels will tolerate it. Farinaceous articles are entirely inappropriate. Limed milk, to which a little gelatine has been added, or rennet whey, may be given ; but in protracted cases, attended with great prostration, and rapid emaciation, the raw meat diet, prepared as recommended by Prof. TROUSSEAU, will be often seized with avidity and well borne. Lean beef or mutton is first finely hashed, pounded in a mortar to a pulp, and then passed through a fine sieve. The thick concentrated juice thus obtained is nutritious and digestible, and, when salted or otherwise flavored, quite acceptable. Give a half to three-quarters of an ounce, in fractional doses the first day. If well borne by the stomach, increase the quantity day by day, until a quarter or half a pound is taken in the course of the twenty-four hours. For the first day or two much of it may pass, hardly changed, from the bowels ; but this alone should not occasion its discontinuance. If too long continued, however, the diet is liable to generate tænia. White of eggs thinned with natural or artificial Seltzer, Vichy, or weak lime water, is an excellent drink, to which a few grains of bicarbonate of soda may be added. Tonics and stimulants are often required. Of the former, minute doses of arsenic, alone, or combined with quinine, or the chloride of iron, or the pernitrate of iron, or the tincture of nux vomica may be given. Wine-whey, or brandy and water, to which a few drops of the aromatic spirits of ammonia have been added, are the best stimulants. The effect of sending the patient to a cool and mountainous region, is immediate and lasting.

PROF. J. LEWIS SMITH, NEW YORK.

461. R. Creasoti, gtt. j.
Aquæ calcis, f.ʒij. M.

Dose—one teaspoonful with a teaspoonful of milk, breast
milk if the infant nurses, repeated p. r. n., for the *vomiting*, so frequent in the summer epidemics of intestinal inflammation in the cities.

This recipe is much used in the Nursery and Child's Hospital, of New York city. Or, the following may be administered:

462. R. Potassæ bicarbonatis, gr.xxv.
Acidi citrici, gr.xvij.
Aquæ amygdale amaræ, f.ʒj.
Aquæ, f.ʒij. M.

Dose—one teaspoonful to a child from eight to twelve months
old, repeated according to the nausea or vomiting.

Another anti-emetic, in these cases, though according to
our author's experience, inferior to those given above, is the
subnitrate of bismuth:

463. R. Bismuthi subnitratis, Əj-ij.
Aquæ menthæ piperitæ,
Syrupi zingiberis, aa. f.ʒj. M.

Dose—one teaspoonful to a child one year old, every hour,
if required for the vomiting.

The bismuth, aside from its sedative effect upon the stomach, also aids in controlling the diarrhœa.

CHOLERA.

DR. FLEMING, QUEEN'S HOSPITAL, BIRMINGHAM.

464. R. Plumbi acetatis, gr.ij.
Liquoris morphiæ acetatis, ℳv.
Acidi acetici diluti, ℳj.
Aquæ destillatæ, f ʒß M.

For one dose every two hours (an hour before or an hour
after food is taken) in a tablespoonful of water. Beef tea
and milk alternately every four hours; warm negus in
moderation.

The value of acetate of lead and opium in diarrhœa is well known. Dr. Fleming calls attention to the *mode of administration*. The astringent action of both lead and morphia is in consequence of their absorption and conveyance in the blood to the affected part. Hence, the marked advantage of giving them in a thorough solution in distilled water, which promotes their passage into the blood. This is further secured by giving the medicine on an empty stomach before meals, and so avoiding, as far as possible, precipitation of the lead by contact with the gastric fluids. In the ordinary lead and opium pill, more or less lead is probably converted into the meconate ; or the pill dissolving slowly in the stomach in contact with the gastric secretions, the lead runs much risk of conversion into the insoluble chloride. One author has adopted this mode of administration for many years, and speaks strongly of its efficiency. In the diarrhœa of children the same mixture, according to the following formula, gives most valuable results.

465. R. Plumbi acetatis,   gr.ss.
          Liquoris morphiæ acetatis, ℳ.ss.
          Acidi acetici diluti,   ℳ.v.
          Aquæ destillatæ,   f.ʒj.   M.

For one dose every five, six or eight hours to a child one
    year of age.

CHRONIC GASTRITIS.

J. M. DACOSTA, M. D , PHILADELPHIA.

466. R. Bismuthi subnitratis,   gr. x-xxx.
          Sodæ bicarbonatis,   gr. x.   M.

For one powder, to be taken after meals, when there is
    acidity.

467. R. Bismuthi subnitratis,   ʒss.
          Acidi hydrocyanici diluti, ℳ.xxiv.
          Misturæ acaciæ,   f.ʒiij.   M.

A teaspoonful after meals.

When there is pyrosis the following is useful :

468. R. Olei succini rectificatæ, f.ℨss.
Misturæ acaciæ, f.ℨijss. M.
Teaspoonful after meals.

GASTRIC ULCER.

J. M. DA COSTA, M. D., PHILADELPHIA.

469. R. Argenti nitratis, gr. v.
Extracti opii, gr. x. M.
For xx pills. One ter die.

CHRONIC TYMPANITIS.

CHARLES MURCHISON, M. D., F. R. S., ETC., LONDON.

Flatulence will often be relieved by the various ethers and the essential oil of peppermint, anise and cajeput, by vegetable charcoal, galbanum and assafœtida. When, however, it is due to decomposition, from deficient or deteriorated bile, those remedies will be found most useful which act by checking decomposition, such as creosote, turpentine or carbolic acid.

470. R. Acidi carbolici fluidi, ℳx-xxx.
Spiritûs chloroformi, ℳx.xxx.
Aquæ menthæ piperitæ, f.ℨss. M.
For one dose.

Or, a pill may be given containing one drop of creosote.*

Flatulence and other dyspeptic symptoms arising from want of bile in the bowels, are also greatly relieved by the use of purified bile from the ox or pig, which may be given in doses of from three to six grains, about two hours after meals. As it is not desirable that the bile should come in contact with the stomach, it is well to give it enclosed in capsules or in pills coated with a solution of tolu in ether.

*Clinical Lectures on Diseases of the Liver, p. 259. Am. ed. 1868.

10

The following recipe will also be found useful for the same purpose:

472. R. Sodæ chloratis, gr. x.
Aquæ menthæ piperitæ, f.℥ss.

For one dose.

CHRONIC HEPATITIS.

J. M. DA COSTA, M. D.

A certain amount of drain should be kept up from the portal circulation. For this purpose, very small doses of podophyllin, cream of tartar, or Rochelle salts, now one and now another, are useful. A very good pill is the following:

472. R. Podophyllin, gr. ss.
Capsici, gr. ⅓.
Pulveris rhei, gr. j. M.

For one pill, to be taken on alternate nights.

In order to reduce the state of induration or enlargement of the liver, the mineral acids may be employed. Or, when the case is not chronic, the salts of soda, as in the following formula:

473. R. Sodæ bicarbonatis, gr. xx.
Infusi gentianæ, f.℥ss. M.

For one dose, three times a day, after or between meals.

Should this fail, employ nitro-muriatic acid baths. Sulphur baths are of advantage; or those of sulphuret of potassium; or the use of sulphur ointment followed by warm baths.

V. DISEASES OF THE URINARY ORGANS.

DIABETES.

WILLIAM AITKEN, M. D., EDINBURGH.

The whole *materia medica* has been exhausted in search of a remedy for this disease. It may be said nearly every patient dies whose treatment is left entirely to drugs. A full and generous diet is unquestionably useful in these cases; but the patient soon gets disgusted with mutton or beef, or both, for breakfast, dinner and supper; he consequently nauseates a meat diet and abandons it altogether. A diet of salt fish has been attempted, but the patient in a short time so loathes it that it has to be given up. A mixed diet, therefore, if contraindicated by some theories, is at least the best to adopt in practice, if only regulated and aided by other means. It will be evident, however, that vegetables which contain a large quantity of saccharine matter should be avoided in some degree, as potatoes, grapes, or other very ripe fruit, and *a fortiori*, sugar itself. That meat and eggs may be taken, if biliary derangement is not induced by them, and fish is a most important article with which to vary the monotony of the dietary. Milk also may be indulged in occasionally, as it is not found that the sugar it contains is readily converted into glucose. Its influence, however, requires watching. It should be combined with half its bulk of lime-water, or to be used in the form of buttermilk. If it be found desirable or necessary to employ alcoholic drinks, a selection should be made from those wines and spirits which are freest from sugar. Of these, clarets may be chemically considered the best, then Burgundy. The so-called "fruit wines" must be interdicted, and of all alcoholic beverages, *weak* brandy and water is the safest. The amount of brandy should always be *measured*. A teaspoonful in a tumblerful of water is generally sufficient for an ordinary dinner drink. No diabetic need expect to recover or continue well who cannot exercise self-control, and make up his mind to be temperate *in all things*. Tea and coffee without sugar may be permitted. If cocoa agree it may be taken prepared *from the nibs only*. When soups are allowed they ought to be really good, and flavored with aromatics or onions to the exclusion of carrots, turnips and peas. They may be thickened with some bran *finely* powdered. Pickles in small quantities may be permitted to convalescents. Lettuces agree well when eaten sparingly with oil and vinegar, or with a little salt only, if the vinegar is likely to disagree.

The great value of *bran cakes* as a substitute for bread in cases of diabetes, has now been established by the experience of so many individuals that its use ought to be insisted upon. The bran should be *thoroughly washed*, so that it may be as free from starch as possible, and *finely powdered*, so that it may not irritate the susceptible mucous membrane of the intestines.

A. BOUCHARDAT, PROF. HYGIENE TO THE FACULTY OF MEDICINE, PARIS, ETC.

Our author has proved by the comparison of the articles of food, which a diabetic may take without prejudice, that it is quite possible to keep up a sufficient degree of variety in the character of the meals. The list of articles which it permits is about as follows: All sorts of meat roasted, stewed and even dressed with spices, but not with flour; fresh water fish and marine fish, in eating which the want of bread is less felt than in eating meat; oysters, muscles, crabs, lobsters, etc.; eggs in all the forms known to the culinary art; rich, good cream, but no milk; of vegetables, spinach, artichokes, asparagus, green beans, the different varieties of cabbage; of the salads, water cresses, endives, lettuce; of fruits, strawberries and peaches. Although the loss of sugar is augmented by a very free supply of liquid, and decreased by a privation of it, yet it is not advisable to forbid the patient to quench his thirst.

THOMAS KING CHAMBERS, M. D., ETC., LONDON.

474. ℞. Pulveris opii,   gr. j.
In pill; to be taken every night.

In some cases opium seems to diminish the secretion of water, and our author has never distinctly traced any harm to its action. Cinchona, however, which, like opium, lessens the flow of urine, has, in Dr. C.'s hands, produced coma and death. He, therefore, shuns it, even when he wishes to give tonics to increase the appetite. He prefers iron and strychnia.

475. ℞. Potassii iodidi,   gr viij.
          Aquæ,   f.ℨss.   M.
For one dose, three times a day.

This drug was prescribed on purely empirical grounds. During its administration the patient continued to gain weight and strength, and did not exhibit any of the usual symptoms of intoxication by iodine.

Patients may drink as much as they feel disposed; for the gratification of the thirst provides the normal outlet for the abnormal constituent of the blood.

J. M. DA COSTA, M. D., PHILADELPHIA.

476. ℞. Pulveris opii,   gr.ss.
For one pill, ter die.

The opium treatment, Dr. DA COSTA has found to be productive of the most good in his hands. He cannot explain the rationale, but he does know that it has the most influence in lessening the thirst, the excretion of sugar and the general waste

of flesh.   The only objection to its use is the risk which is run
of accustoming the patient to taking this powerful drug.

477. R. Potassæ permanganatis,   gr.j.
           Aquæ destillatæ,   f.ʒij.   M.
For one dose, ter die.

The above remedy Dr. Da C. employed a few years ago,
but not with very satisfactory results.

478. R. Ammoniæ carbonatis,   gr.x–xx.
           Aquæ cinnamomi,   f.ʒss.   M.
For one dose three or four times a day.   This, in addition
   to a moderately restricted diet, forms a very good treatment.

479. R. Tr. ferri chloridi,   gtt.xx-lx.
For one dose, in water, ter die.

This will often do good when nothing else proves of service.
The pepsin and rennet treatments amount to nothing.   Dr.
Da Costa has tried them faithfully and merely lost time.

Diet.—Although substances containing a great deal of
sugar, and still more those containing a great deal of starch—
which is readily converted into sugar—are poisonous to diabetic patients, yet it is not advisable to put them on a strictly
animal diet.   Such a regimen is irksome in the extreme, and
will not be followed out.   After all it is not the saccharine
substances which go into the body that cause diabetes; they
merely add to it.   It is simply impossible to avoid giving
food capable of being converted into sugar.

Bran-bread is perfectly unobjectionable; but a small quantity of ordinary bread should be permitted.   Cakes made
from almond-flour are favorites with English physicians.
Fruits should be interdicted and potatoes sparingly used.

Bran baths (two or three handfuls of bran in the bath) are
very serviceable in relieving the dryness of the skin, which is
so troublesome.

480. ℞. Infusi Cascarillæ,   Oj.

A wine-glassful three or four times a day.

The use of this light bitter lessens the thirst. Of course, water must be allowed beside ; but the infusion will take the place of a good deal of water.

THOMAS HAWKES TANNER, M. D., F. L. S., LONDON, ETC.

481. ℞. Ferri ammoniæ citratis,   ʒj.
       Spr. ammoniæ aromatici,   f.ʒvj.
       Potassæ bicarbonatis,   ʒij.
       Infusi calumbæ,   q. s. ad, f.ʒiij.   M.

A tablespoonful to be taken twice a day, with one table-spoonful of lemon juice in a little water.

This recipe often proves very valuable. It should be taken for two or three weeks at a time, then discontinued, and recommended according to the general strength.

482. ℞. Pulveris opii,
       Quiniæ sulphatis,   aa. gr. j.
For one pill, ter die. Useful in some cases.

483. ℞. Creasoti,   ♏xx.
       Pulveris aromatici,   ℈iv.
       Mucilaginis acaciæ,   q. s.   M.
    Fiat massa, in pilulas xx dividenda.
One to be taken three times a day.

It is believed that this remedy tends to check the conversion of the food into sugar.

BRIGHT'S DISEASE.

WILLIAM AITKEN, M. D., EDIN.

484. ℞. Acidi gallici,   ʒj-ij.
       Acidi sulphurici diluti,   f.ʒss.
       Tincturæ lupuli,   f.ʒj.
       Infusi lupuli,   f.ʒvj.   M.

A tablespoonful ter die, when the urine is "smoky" or when blood is seen on microscopic examination.

The objection to these remedies is the constipation they are apt to induce. Two or three movements from the bowels should be secured daily. The medicine most generally useful is

485. ℞. Pulveris jalapæ compositæ, ʒss–j.
Pulveris zingiberis, gr.ij. M.

For one dose, to be taken in the morning, fasting, in a wine-glass of water.

When by repetition this loses its effect, and elaterium be not deemed advisable, the following saline draught may prove efficient:

486. ℞. Magnesiæ sulphatis,
vel
Sodæ sulphatis, ʒj–ij.
Ætheris sulphurici, ♏ x.
Acidi sulphurici diluti, ♏ x.
Ferri sulphatis, gr. j–ij.
Aquæ menthæ viridis, f.ʒiij–iv. M.

For one dose; to be taken the first thing in the morning, once or twice a week. It ought to produce two or three loose and watery evacuations.

When dyspeptic symptoms predominate, the following pill is useful, (recommended by Dr. GOODFELLOW):

487. ℞. Ferri sulphatis, gr.j.
Extracti nucis vomicæ, gr. ss–j.
Pilulæ galbani compositæ, gr. ij–iij. M.

For one pill; to be taken twice or thrice daily.

If there be co-existent bronchitis

488. ℞. Spiritûs ætheris nitrosi, f.ʒjss–iij.
Oxymellis scillæ, f.ʒss.
Liquoris ammoniæ acetatis, f.ʒiv. M.

A tablespoonful to be taken immediately with the above pills. If much nausea prevail, add to the prescription
Acidi hydrocyanici diluti, ♏xxiv.

And apply mustard to the stomach.

For the want of sleep often complained of, henbane should be given instead of opium, which will constipate.

If diarrhœa occur, it must not be too suddenly checked. Order

489. ℞. Spiritûs ammoniæ aromat.,
Tincturæ kino.   aa f.℥j   M.

A teaspoonful in water after every loose stool.

This in general will be all that will be found necessary.

Mercurial preparations are dangerous in Bright's disease, because of the tendency to salivation. The following may be used instead :

490. ℞. Podophyllin,   gr. ¼-j.
Extracti hyoscyami,   gr. j.   M.

For one pill.

The following is also serviceable as an occasional mild aperient pill :

491. ℞. Pilulæ rhei compositæ,   gr ij-iij.
Extracti nucis vomicæ,   gr.j
Pilulæ galbani compositæ,   gr.ij.   M.

For one pill.

THOMAS KING CHAMBERS, M. D., ETC., LONDON.

492. ℞. Tincturæ ferri chloridi,   f ℥ij.
Potassæ nitratis,   ℨij.
Aquæ camphoræ,   f.℥iv.   M.

A tablespoonful in water, ter die, and a hot-air bath every night.*

J. M. DA COSTA, M. D.

For the treatment of an *acute* case, following scarlatina :

493. ℞. Pulv. ipecac. com.   gr.iij.
Potassæ nitratis,   gr. v.   M.

For one powder to be taken at night. And

494. ℞. Liquoris ammoniæ acetatis, f.℥iij.
Spiritûs ætheris nitrosi,   ℳxx.
Syrupi tolutanus.   ℳxxxvij.
Tincturæ digitalis,   ℳiij.   M.

For one dose ter die.

The skin should be made to act freely by means of hot baths and a few drachms of Rochelle salts administered on

---

*The Renewal of Life. Am. Ed. p.465.

alternate days. The diet should be mild and unirritating, all stimulants being avoided.

495. R. Tincturæ ferri chloridi, f ʒss.
Acidi acetici, f.ʒss.
Mix and add
Liq. ammoniæ acetatis, f.ʒv.
Curaçoæ, f.ʒij. M.
Tablespoonful ter die.

This recipe, a modification of *Basham's* mixture, is useful in chronic Bright's disease, and in all those cases in which the union of a tonic and diuretic effect is desired.

DR. FELIX VON NIEMEYER, PROFESSOR UNIVERSITY OF TUBINGEN.

The loss of albumen from the blood being the immediate cause of most of the symptoms of the disease, the most important task of the physician is to cover the loss of albumen by a diet rich in protein substances and by appropriate medication. Soft boiled eggs, milk, strong meat broths and roast beef, in as large quantity as the patient is able to digest, are probably the best preventives of the dropsy. Besides this a moderate quantity of beer or good wine should be prescribed, as by the use of these the waste of tissue is retarded and nutrition prevented. Quinine and iron are the most suitable medicines. Our author has obtained most brilliant results where all other treatment had failed, by putting the patients upon an exclusive diet of milk. They took no medicine whatever, but drank five or six pints of cow's milk daily. After the "cure" had been continued in this manner for about five weeks some of the patients, who, prior to the treatment, had been in the most wretched condition, had got rid of their dropsy, recovered an appearance of health and regained so much strength as even to be able to perform hard labor.

If the above measures fail in averting or allaying the dropsy, active diaphoresis is strongly to be recommended. Patients in an advanced state of dropsy often rid themselves of it completely in a few weeks by the daily use of a hot bath, of a temperature of 80° to 100° F., followed by sweating for two hours in woolen blankets. Debilitated patients sometimes, however, suffer so much from this treatment as to compel its discontinuance.

Whatever the theoretical objections to the employment of diuretics may be, yet, in desperate cases, recourse should always be had to them. Squills and other stimulating diuretics must not be employed without the utmost caution. But there are certain salts, particularly cream of tartar, which are decidedly beneficial in their effect. The free use of buttermilk, conjoined with the employment of cream of tartar and small doses of Dover's powder will prove serviceable.

The drastic cathartics should only be resorted to in cases of extreme need, since the patient is liable to be affected by them, and since, by their persistent use, the digestion becomes impaired. Those most frequently employed are colocynth and

496. R. Elaterii, gr. 1-6-½.
For one dose, pro re rata.

THOMAS HAWKES TANNER, M. D., F. L. S. ETC., LONDON.

497. R Salacini, ℥ij.
Glycerinæ, f.℥j.
Tincturæ aurantii corticis, f.℥ij. M.
A teaspoonful in a wineglassful of water night and morning·
Often useful in imparting a temporary sense of renovation·
So also is

498. R. Ferri et ammoniæ citratis, ℥ss.
Spiritûs vini gallici, f.℥j.
Vini pepsinæ, f.℥ss.
Aquæ, f.℥ijss. M.
One-half to be taken every day at dinner.

For the administration of elaterium, when indicated, our author employs the following formulæ:

499. R. Elaterii,   gr.jss.
      Pulveris capsici,   gr.lx.
      Hydrargyri chloridi mitis, gr.xij.
      . Extracti hyoscyami,   gr.xviij.   M.

For xij pills. Two to be taken at a dose.

The capsicum prevents the nausea which the elaterium often produces. If a very active purgative is required, the dose may be doubled.

500. R. Elaterii,   gr.j.
      Spiritûs ætheris nitrosi,   f.Ʒss.
      Liquoris ammoniæ acetatis, f.Ʒj.
      Syrupi zingiberis,   f.Ʒiij.   M.

One small teaspoonful in a wineglassful of water every two hours, until the bowels are freely acted on.

SABULOUS DEPOSITS IN THE BLADDER.

S. W. BUTLER, M. D., PHILADELPHIA.

501. R. Fresh root of hydrangea
        arborescens,   2 lbs.
      Water,   6 qts.

Boil down to two quarts, strain, and add one quart of honey and boil down to one quart.

Dose—A teaspoonful twice or three times a day.

Dr. BUTLER highly recommends this remedy in cases of sabulous and gravelly deposits in the bladder. Under its use large quantities of sand and gravel have been removed.

A fluid extract of the hydrangea arborescens is prepared by the leading pharmaceutists, and may be readily obtained.*

---

*Medical and Surgical Reporter. Nov. 9th, 1861. p. 143.

DROPSY FOLLOWING SCARLATINA.

J. M. DA COSTA, M. D., PHILADELPHIA.

In cases of dropsy following scarlatina, rest in bed should be insisted upon. The diet should be regulated ; milk, beef-tea, chicken broth, oysters, and other easily digested food being alone proper. No articles of diet, which, by leaving matter difficult of excretion will throw work upon the kidneys, should be allowed.

As regards medical treatment, bearing in mind the fact that the kidneys are in a state of active engorgement, unirritating diuretics should be prescribed. For instance, digitalis comes admirably into play.

502 R. Tincturæ digitalis,   ℳ xxiv.
        Liquoris potassæ citratis,   f.ℨiv.   M.
     A tablespoonful ter die.

Dry cups over the kidneys may be ordered. In cases where there is a decided amount of albumen, and an appearance of blood in the urine even wet cups may be applied with good effect.

———

ASCITES.

C. MURCHISON, M. D., F. R. S., ETC., LONDON.

503. R. Pulveris scillæ,   gr.jss.
        Pulveris digitalis,   gr.ss.
        Pilulæ hydrargyri,   gr.ij.   M.
     For one pill, two or three times a day.

This is a pill which has enjoyed a long and merited reputation for treatment of dropsy in the Middlesex Hospital.

Diuresis will also sometimes be induced by fomenting the abdomen with an infusion of digitalis of about four times the usual strength.

504. R. Potassæ acetatis,   gr. xx.
      Spiritûs ætheris nitrosi,   f.ʒss.
      Decocti scoparii,   f.ʒjss.   M.

To be given with each dose of the above pills.

At the same time an ointment composed of equal parts of blue ointment and ointment of belladonna is to be applied over the abdomen. This treatment is a successful one in cases of acites due to cirrhosis.

VI. DISEASES OF THE BLOOD.

RHEUMATISM.

WILLIAM AITKEN, M. D., EDINBURGH.

505. R. Hydrargyri chlo. mitis,   gr.v.
      Pulveris jalapæ comp.,   Əij-ʒj.
      Pulveris zingiberis,   gr.iij-v.   M.

This powder to be taken in a little milk at the commencement of rheumatic fever. It will secure an abundant secretion from the glandular follicles of the intestinal mucous membrane. The daily evacuation of the bowels is then to be maintained by salines, such as Rochelle or Epsom salts.

506. R. Veratriæ,
      Extracti opii,   aa. gr.j.   M.

For ten pills.

Take two the first day, three the second, four the third, five the fourth, and so on, increasing one pill each day, until the condition of the pulse or the irritation of the mucous membrane compels a diminution.

506. R. Pulveris guaiaci,   ʒj.
      Pulveris rhei,   ʒij.
      Potassæ bitartratis,
      Sulphuris sublim.,   aa. ʒj.
      Pulv. nucis moschatæ,   ʒij.
      Mellis,   ℔j.

Misce bene.

Of this compound, two large spoonfuls are to be taken night and morning.

It is used in some large hospitals, both civil and military, in the treatment of old chronic cases. The beneficial effects of guaiacum are obtained in those cases which are unaccompanied by perspiration, and in which the excreting organs are gently excited by this drug. When, however, there is already free diaphoresis, and when it neither purges nor acts as a diuretic, very little benefit may be expected from it.

THOMAS KING CHAMBERS, M. D., ETC., LONDON.

Our author calls rheumatic fever "a pleasant disease for the doctor to treat, though not for the patient to bear," and gives a very simple, uniform plan of treatment, which, he states, hardly ever requires modification.

Bedding.—The patient's bed is made in a peculiar fashion. No linen should touch the skin. A slight calico shift or shirt may be allowed; but if the patients possess underclothing only of the prohibited sort, they are better naked. Even a linen front to the shirt is dangerous. The sheets should be removed and the body carefully wrapped in blankets, the newest and fluffiest that can be got. The head is to be carefully protected from currents of air.

Fomentations.—Those joints or limbs which are swollen, red, or painful, are to be wrapped up in flannels, soaked either in hot water, or in a decoction of poppy heads, with half an ounce of carbonate of soda to each pint.

Curative Drugs.—If the skin is red, swollen and painful about the joints, if motion is impossible or the cause of exquisite suffering, and especially if these phenomena are metastatic, then the "alkaline treatment" is employed, as follows :

508. R. Potassæ bicarbonatis, Əj.
Aquæ camphoræ, f.ʒij. M.
For one dose, to be repeated every three hours, day or night, when awake.

If, however, the above symptoms are insignificant, and the pain is felt more in the bones, being intensified by pressure rather than by motion, and fixed, not metastatic, then two grains of iodide of potassium are to be added to each dose. So soon as the symptoms take a favorable turn, the alkali is to be omitted altogether and only the iodide of potassium given.

Palliatives.—Opium is to be administered in amounts proportionate to the subjective sensation of pain—from one to two grains at a dose. Immediately upon the relief of the pain the quantity is diminished.

Pure opium in the pill, and the tincture, are the best preparations. If the pain remains fixed in one joint after it has left the other places, leeches are to be applied there and the part kept poulticed. Bruised laurel leaves may be mixed with the poultice. If the heart become affected, leeches and poultices are to be applied to the cardiac region.

Diet.—The food is to be varied to some extent by the social and personal state of the patients. If they have been hearty and well-to-do persons before the attack, simple diet is proper, *i. e.*, bread and butter, gruel and tea. If they have been ill-nourished, a pint of broth or of beef-tea is added. Meat, even during convalescence, often does harm, seeming to turn into lactic acid. Vegetable food should be pretty closely adhered to in order to avoid a relapse.

J. M. DA COSTA, M. D., PHILADELPHIA.

509. R . Ammonii bromidi, $\mathbb{3}$ss.
          Tincturæ aurantii corticis,   f.$\mathbb{3}$ss.
          Aquæ,   f.$\mathbb{3}$ijss.   M.

A dessertspoonful every three hours, excepting at night. In acute rheumatic fever the beneficial influence of the bromide of ammonium is undoubted.

Over the pains and aches of chronic rheumatism this remedy also exerts an unquestionable control ; but in chronic
rheumatism it is decidedly inferior to iodide of potassium.
But slight amelioration follows its use in persistent swelling
of the joints of rheumatic origin and none in rheumatism due
to a venereal taint.

HYPODERMIC INJECTION.

510. R. Atropiæ sulphatis, gr.1-50.
Aquæ destillatæ, ℥x. M.

For one injection, to be thrown under the skin, in certain
forms of muscular rheumatism, particularly wry neck,
over or into the rigid parts. To be repeated once a day.
Sometimes the most marked and speedy relief follows
this treatment, after ordinary remedies have failed.

511. R. Potassii iodidi, ʒj.
Vini colchici radicis, f.ʒij.
Morphiæ sulphatis, gr.iij.
Syrupi, f.ʒj.
Aquæ, f.ʒij. M.

A teaspoonful three times a day, after meals, in muscular
rheumatism.

512. R. Potassæ carbonatis, gr.x.
Potassæ acetatis, gr.xv.
Vini colchici seminis, ℥xx.
Syrupi tolutanus, f ʒss.
Aquæ, ℥x. M.

For one dose ter die, in lumbago, the following liniment to
be rubbed in night and morning :

513. R. Chloroformi, f.ʒss.
Linimenti ammoniæ, f.ʒjss.
Linimenti saponis, f.ʒij. M.

In a week or ten days, after the pain has subsided, apply

514. R. Emplastri picis cum cantharide,
Emplastri Logani, aa. equal parts. M.

The quickest way of treating lumbago is by throwing 1-50th
of a grain of atropia under the skin near the affected muscles.
This will sometimes dissipate the attack as if by magic.

11

515. R.   Tincturæ guaiaci,   f.ℨij.
          Vini colchici radicis,   f.ℨij.
          Potassæ bicarbonatis,   ℨss.
          Syrupi aurantii corticis,   f.ℨij·   M.

A dessertspoonful in water, ter die, in *rheumatic arthritis.*
Also have the joints wrapped in cotton, and painted,
morning and evening, with equal parts of tincture of
iodine and alcohol until they become sore. *Sulphur baths*
are beneficial. Small blisters are also of service in the
neighborhood of the joints, frequently repeated and
dressed with

516. R.   Morphiæ acetatis,   gr.ᴊ.
          Pulveris marantæ,   gr.j   M.

Arsenic is an excellent remedy in rheumatic arthritis. It
may be administered as follows :

517. R.   Liquoris potassæ arsenitis,   f.ℨij.
          Potassii iodidi,   ℨij.
          Syrupi,   f.ℨiij.   M.
A teaspoonful, ter die, between meals, in water.

518. R.   Vini colchici seminis,   f.ℨj.
          Potassæ acetatis,   ℨvj.
          Spiritûs ætheris nitrosi,
          Syrupi,   aa. f.ℨj.   M.
A teaspoonful ter die in *pleurodynia.*

Also use a stimulating liniment and keep the parts **warm.**

519. R.   Potassii iodidi,   ℨij.
          Vini colchici radicis,   f.ℨss.
          Extracti cinchonæ fluidi,
          Elixir cinchonæ,   aa f.ℨjss.   M.
A dessertspoonful ter die in muscular rheumatism.

Cinchona disguises the taste of the iodide of potassium.
Quinine is often serviceable alone in those cases in which
the joints are not affected, and in which there is pallor and
loss of strength.

520. R.   Potassii iodidi,   ℨij.
          Tincturæ belladonnæ,   f.ℨiss.
          Syrupi aurantii corticis,   f.ℨss.
          Aquæ,   q. s. ad. f.ℨiij.   M.
A teaspoonful ter die in muscular rheumatism associated with
vesical incontinence.

In case of swelling and stiffness of the joints following acute rheumatism, nothing does so much good as sulphur baths. When practicable, send the patient to the sulphur baths of Virginia to get rid of these remnants of the acute attack. Artificial sulphur baths are also useful. When these are not practicable great benefit will be obtained from the iodide of potassium and compound tincture of guiaici.

521. R. Potassii iodidi,   ʒij.
           Tincturæ guiaici comp., f.ʒvj.   M.
      A dessertspoonful ter die.

This prescription is a very efficient though not an elegant one. It is very unpleasant to take. There are few who will go on with it for a long enough time to get its full therapeutical effects.

In the treatment of the *sub-acute form of muscular rheumatism*, nitrate of potash is very valuable. This is an old remedy, but the advance of therapeutics has shown no other, in this form of rheumatism, of which our author thinks so highly. It may be conjoined with Dover's powder.

522. R. Potassæ nitratis,   gr.xv.
           Pulv. ipecacuanhæ comp., gr.iij.   M
      For one powder, to be taken every fourth hour.

PROF. AUSTIN FLINT, NEW YORK.

The *tincture of aconite* applied to the parts in *acute articular rheumatism* affords marked relief. Extension of the limbs, by means of an apparatus for that purpose, or by simply a cord, pulley and weight, in order to relieve the pressure of the articular surfaces upon each other, has been tried with success, in some cases, at Bellevue Hospital.

C. HANDFIELD JONES, M. B., CANTAB, ETC., LONDON.

523. R. Potassii iodidi,   gr. xxxvj.
          Ammoniæ muriatis,   ℨiij.
          Vini colchici radicis,   f.ℨj.
          Tincturæ opii,   ♏l.
          Infusi gentianæ comp.,   f.ℨvj.   M.
      A tablespoonful, ter die, in *muscular rheumatism.*

HENRY POWER, F. R. C. S., ETC., LONDON.

524. R. Tinct. aconiti radicis,   ♏xxiv.
          Tincturæ colchici,   f.ℨijss.
          Aquæ camphoræ,   f.ℨiv.   M.
      A tablespoonful three or four times a day.

In feeble patients, in women, and in children of from 10 to
14 years of age, the quantity of each tincture may be reduced
to one-half; and when the water is, or has been, high colored,
with deposits of the lithates, the bicarbonate and nitrate of
potash, with a little spirit of chloroform or of nitric acid may
be advantageously added.

Our author has employed these remedies to a considerable
extent in various cases of rheumatic disease for the last few
years, and finds their use attended with great benefit when judiciously employed. They are particularly beneficial in *rheumatic
ophthalmia.*

In many instances when the disease presents a periodic
character, recurring with great regularity at a certain period
of the night, the addition of two or three grains of quinine, or
of a few minims of Fowler's solution, may be made.

The only two unpleasant symptoms our author has ever
observed have been nausea and sickness, probably produced
by the colchicum, and tingling, and numbness of the fingers,
proceeding from an overdose of the aconite. These, however,
have been rare and exceptional inconveniences, and have, in
every instance, promptly disappeared with the intermission of
the medicine.*

---

*The *Practitioner*, London, October, 1869, p. 227.

THOMAS HAWKES TANNER, M. D., F. L. S., ETC., LONDON.

525. R. Ammoniæ muriatis,   ℥iij.
          Liquoris ammoniæ acetatis, f.℥vj.
          Aquæ camphoræ,   q. s. ad. f.℥vj.   M.

A tablespoonful, in water, every four hours. Useful in some
varieties of rheumatism in which the fibrin of the blood
is in excess. The efficacy of this prescription is increased
by giving two drachms of cream of tartar, in half a pint
of water, early in the morning.

TREATMENT OF ACUTE RHEUMATISM

IN THE HOSPITALS OF GREAT BRITAIN.

In the *British Medical Journal* for January 2, 1869,
p. 8, there is recorded the practice of the principal London Hospitals in this troublesome affection.

The treatment of acute rheumatism is a subject which has
received, perhaps, of late years, greater attention from the
profession than any other disease. The numerous remedies
which have been enthusiastically brought forward as specifics,
have received a fair and very extensive trial, and as a result
have all sunk to a level far below the expectations of their
advocates. They received a severe blow after the experience
of Dr. GULL, Dr. A. P. STEWART, and others had been made
known, in treating the diseases by the expectant method. It
was then shown that acute rheumatism tended to recover
without medicinal treatment of any kind, and that not a
few cases recovered under circumstances almost as favorable
as those treated by the received alkaline remedies. The aggregate results of expectant treatment are generally recog-
nized to be less favorable than the remedial, both as to the

length of the acute stage and the tendency to recurrence and
chronic affections of the joints. Yet, after reading the following notes from men of large experience and of acknowledged
authority, in which the marked difference of opinion as to
the remedy or remedies, and even the doses of these to be
employed, is so apparent, one cannot but feel how uncertain
our therapeutical knowledge is on the subject, and ask the
question : What should be the treatment of acute rheumatism? Of the various remedies which are in use at present,
the alkaline seems to have gained the greatest share of confidence at the hands of the profession; still the different success which has attended the practice of one physician with
another, is a matter which requires solution. That acute
rheumatism varies in severity from year to year, and in different localities, is most likely, and perhaps much may be
explained in this way as to the various opinions arrived at by
leading authorities on the subject, but it is palpable that much
has still to be learnt before we can speak dogmatically, as
men have been apt hitherto to do, in the treatment of this
painful and mischievous malady.

Guy's Hospital.—Dr. WILKES considers rheumatic fever
one of the most difficult diseases for study, the principal
questions for elucidation being its natural progress ; the most
usual time for heart-implication; whether any remedies will
cut short the disease ; and, if so, whether its curtailment prevents the cardiac affection. Dr. Wilkes says that individual
practitioners will answer such questions most positively, but
the profession is by no means agreed about their solution.
Although he entertains a doubt as to the best method of treatment, he can fully endorse the statements made by the enthu-
siastic supporters of each particular remedy. He has fully
tried them all, and has seen patients rapidly recover under

the use of lemon-juice with a diminution of the pulse ; he has
had several cases where the blister treatment has been followed
by the most marked success ; he has given salines in large
doses, and with a speedy good result ; and he has adopted
the eliminative plan of wrapping the patient in a blan-
ket with the most marked benefit.   His difficulty is, not that
these plans always fail, for then they might forever be put
aside, but they appear to be eminently successful.   This is the
good side of the case ; the other picture is, that they all sometimes fail ; the patient lingers on week after week, sometimes
with the articular affection alone, and sometimes with the
chest-complication.   He had endeavored, also, by the most
careful analysis of symptoms, to discover the connection between the time of the occurrence of the alkanlinity of the
urine and the abatement of the symptoms, but without success, although very often the doctrine usually taught that the
symptoms depart when the urine becomes alkaline, is found to
be true.   It was owing to this uncertainty as to the best mode
of treatment, that the physicians at Guy's determined to study
the disease uncomplicated by remedies ; for it was assuredly
true that no one knew what course rheumatic fever might
take if left alone, or at least no one was supposed to know.

Dr. Wilkes believes that the treatment almost universally
adopted until quite lately was equivalent to the "do nothing"
system ; that the few grains of saline three or four times a
day could have had no influence in checking the disease.   The
medical man was well satisfied with himself because the man
did not die, forgetting that rheumatism, with its complications
is rarely a fatal disease.   If a patient with this disease took
the usual saline, and then had endocarditis, pericarditis, and
pleurisy, and, after weeks in bed, escaped with his life, friends,
patient, and doctor congratulated one another on the favorable

termination. The only correct view which can be taken, would be that by a medical patient himself; he would consider that his case had ended well if he escaped without cardiac disease, and badly indeed if he rose from his bed with an affection of the heart. He would say to his adviser, cut the disease short if this will arrest its progress to my heart, but by all means let it run its course if by this means my heart will escape. Dr. Wilkes thinks that it is by no means yet determined that absence of articular symptoms implies the less liability to cardiac disease, or that, after the administration of those remedies which are supposed to shorten the disease, the heart has more readily escaped. He has had lately two patients who were taking half a drachm of bicarbonate of potash every three hours, and very soon the urine became alkaline; after about two days the pains in the joints were lessened, but on the third day acute pericarditis appeared in both patients. In three other cases, now in the clinical ward, one is on quinine, another in blankets, and a third on the blister treatment; although the disease has been protracted, no cardiac affection has appeared in any of them. He thinks, however, that the two cases on the alkaline treatment were unusual, not on account of any antidotal influence excited by the medicine, but from the fact of the pericarditis occurring after the patient had been admitted. He thinks, that in most hospital cases the patient is never taken in until very ill, and then the cardiac inflammation already exists if about to occur at all; if not then present, it is not likely to happen under any treatment. He would not speak dogmatically on this subject, but believes it to be in the main true; and, if so, it is only in private cases, where the patient is seen early, that the definite action of the remedies can be discovered. He would also insist on a fact, too often overlooked, that, whilst

lymph on a pericardial surface at once makes itself known by
the *frottement*, an equivalent inflammation of the endocardial
surface might present no altered sound, and it would be only
by the gradual thickening which subsequently takes place,
that the fact would be known.   This was the case of a man
who had acute pericarditis, and it was a question whether
there was also an endocardial sound synchronous with the
rub; when the latter passed off, however, no murmur refer-
able to the valves could be heard, nor at any subsequent period
whilst the patient was in bed.   After two or three weeks he
was discharged; but before his leaving, the stethoscope was
placed over the heart, when a distinct systolic murmur was
heard; this had been more than eight weeks developing.

Dr. Wilkes considers that the main point for consideration
is the discovery of that treatment which will bring the patient
through without implication of the heart; that the profession
has not yet arrived at this, is certain from the fact of the
thousands who die annually of cardiac disease having its
origin in rheumatism. If treating a private patient, and having sufficient faith in orthodox remedies until they can be
superseded, he prescribes the acetate and nitrate of potash
with an opiate at night, occasional blisters to the joints to relieve pain, with flannel next to the skin, etc.   Since these
notes were written, Dr. Wilkes has published an interesting
paper in the *Practitioner*, recommending the use of tincture
of aconite, which he has found of marked benefit, given in
frequently repeated small doses, in several cases of acute
rheumatism.

St. George's Hospital.—The treatment of rheumatic fever
adopted by Dr. FULLER is essentially alkaline, and consists
not only in administering salines and small doses of alkalies,
but in pushing alkalies as rapidly as possible to the point of

producing alkalinity of the secretions. When a patient is admitted into the hospital, Dr. Fuller's first care is to determine that the disease under which he is suffering is really rheumatic fever; for Dr. Fuller maintains, and acts upon the belief, that cases of rheumatic gout in its acute stage simulate, and are often mistaken for rheumatic fever. He asserts, indeed, that a want of discrimination as to the true nature of the disease is one of the causes of the failure which some persons experience in their treatment of so-called rheumatic fever by alkalies—true rheumatic fever yielding readily to these remedies, which exercise little control over acute rheumatic gout. The points of distinction on which Dr. Fuller mostly relies as indicating rheumatic gout are : 1. The complexion of the patient, which is seldom so pallid as in rheumatic fever. 2. The state of the skin, which is more inelastic and doughy than in rheumatic fever. 3. The existence of perspiration devoid of a strongly marked rheumatic odor. 4. Tendency to swelling in the small joints of the hands. 5. The comparative absence of redness and coating of the tongue. 6. The absence of a copious deposit of lithates in the urine. When these conditions co-exist, Dr. Fuller disregards the heat, pain, redness, and swelling of the joints, orders the patient out of bed, prescribes a cold shower-bath, and gives a generous diet, including meat and porter. At the same time, he administers bark and the mineral acids with cod-liver oil, or the mineral acids with tincture of iodine and cod-liver oil ; or, if the urine be high-colored, acid, and somewhat turbid, quinine, or strychnia dissolved in citric acid, and given in effervescence with a dose of the bicarbonate of potash or soda, the secretions being regulated by an occasional alterative or a dose of some mild aperient. When, on the other hand, the case is manifestly true rheumatic fever, the alkalies—whether potash or soda,

appears immaterial, and Dr. Fuller often combines the two—
are given to the extent of two drachms every three or four
hours until the urine is rendered alkaline. Dr. Fuller usually
prescribes

526. R. Sodæ bicarbonatis, ʒjss.
Potassæ acetatis, ʒss.
Liquoris ammoniæ acetatis, f.ʒiij.
Aquæ, f.ʒjss. M.

For one dose; to be taken in a state of effervescence
in combination with

527. R. Acidi citrici, ʒss.
Aquæ, f.ʒij. M.

The quantity of the fluid not only takes off from the nauseous quality of the medicine, but promotes its absorption, and
thereby facilitates its action. As soon as the urine manifests
an alkaline reaction, the dose is repeated three times only in
twenty-four hours; and on the following day, if the urine still
remains alkaline, twice only. After three days, two doses only
of this mixture in twenty-four hours usually suffice to keep
the urine alkaline; and then Dr. Fuller adds two grains of
quinine to each dose; or, if quinine be not well borne, he substitutes

528. R. Sodæ bicarbonatis,
Potassæ acetatis, aa. ʒss.
Tincturæ cinchonæ, f.ʒiss.
Decocti cinchonæ flavæ, f.ʒjss. M.

For one dose.

Day by day, as the tongue cleans, and the other symptoms
subside, the quantity of alkali is cautiously diminished until a
simple quinine draught is taken; but the condition of the
urine is constantly watched with a view to the immediate administration of a small quantity of alkali, should the least
acidity call for its use. The diet is another point on which
Dr. Fuller lays particular stress; he insists that strict absti-

nence from solid food is of far more importance than in gout. He gives beef tea or broth throughout, and, if stimulants appear to be needed, does not object to their being administered cautiously; but he withholds solid food until the tongue is quite clean, and has often proved to his class in the wards that a too early recourse to meat will induce a relapse, and prevent convalescence. Under this treatment, Dr. Fuller says, the pains commonly subside in five or six days, and the patients are seldom ten days in bed. Moreover, the heart may be regarded as safe from attack; for in two instances only, in the whole course of his hospital experience, has inflammation either of the endocardium or pericardium arisen after the patient has been twenty-four hours under treatment, and in one of these cases the alkalies had been imprudently abandoned under the belief that the patient was convalescent.

The treatment employed by Dr. BARCLAY is also alkaline. He follows out this plan after a long experience, believing that alkalies diminish the duration and the pain of the acute stage, and that, by maintaining the alkalescence of the secretions the disease is less liable to become chronic. It prevents, almost invariably, he believes, inflammation of the heart after the system has been fully brought under the influence of the alkali. Dr. Barclay also gives mercury to improve and correct the excretions, and opium in modified doses, to afford, when necessary, desired rest.

Edinburgh Royal Infirmary.—In the clinical wards of this infirmary the following treatment is adopted by Dr. Laycock. The patient is put to bed with flannel next the skin, in order to favor and absorb perspiration and prevent sudden chill. When there is great pain and sleeplessness, a full dose of Dover's powder is administered. If the skin be hot and dry, a hot vapor-bath is resorted to with advantage. In cases of biliary derangement and constipation the patient is freely purged. Calomel is generally given. The principal treatment consists in the administration of carbonate or nitrate of potash in drachm doses, every three or four hours, which Dr. Laycock finds greatly to alleviate the suffering, and considers beneficial in promoting the elimination of irritating substances from the economy. The treatment is the usual diathetic treatment of rheumatic affections in whatever form they occur. When the rheumatic affection is of the the bursal form, colchicum is combined with the alkaline remedies. Quinine is given in those cases characterized by great irritability and restlessness, with marked benefit. Opiate, alkaline, or nitrate of potash epithems are applied to the affected joints. Blisters are not generally employed, except in cases of long continued pain in particular joints. In such cases they seem to produce marked relief in a short time. Lung complications, such as pleurisy, pneumonia, etc.. are treated diathetically by alkalies, and by the local application of opiate or alkaline epithems over, or a blister near, the seat of pain. If a heart affection be established, the treatment consists in :

529. R. Pulveris opii,
          Hydrargyri chlor. mitis, aa. gr. ¼.   M.
    For one pill, to be given every two or three hours.

Heart affections are not so liable to come on when the treatment by alkalies and the wrapping-up in blankets has been adopted at the commencement of the disease. The cases of heart complications are chiefly those where the affection was established before admission into the hospital.

In the *British Medical Journal* for January 9, 1869, p. 27, the report on the practice of the Hospitals of Great Britain in Acute Rheumatism, is continued.

St. Bartholomew's Hospital.—Dr. FARRE's ordinary mode

of treating acute rheumatism is the "alkaline." He usually commences with three five-grain doses of calomel, followed by

530. ℞. Magnesiæ sulphatis, ℨij.
Tincturæ sennæ, f.ℨij.
Spiritûs ammoniæ aromat., ℳxx.
Infusi sennæ compositi, f.ℨjss. M.

For one dose.

He repeats this daily till the evacuations are natural. At the same time he gives the bicarbonate or acetate of potash, in twenty or thirty-grain doses, every four or six hours, according to the severity of the attack, generally using the former, but preferring the latter when there is synovial effusion.

When there is little or no perspiration, or when the heart is much excited, he adds ten or fifteen grains of nitrate of potash. He not unfrequently, also, gives one grain of opium every night. For local treatment, especially of the smaller joints, he relies chiefly on the tincture or liniment of iodine, using the tincture for women and children, the liniment for the robust; one or the other of these is used in almost every case, and with nearly certain relief. When, however, there is synovial effusion, Dr. Farre uses either mustard or cantharides plaster. Mustard is always useful, especially when applied to the larger joints, as the shoulder. The appetite being always faulty, Dr. Farre gives milk diet until the tongue is clean or cleaning. Meat given before it can be digested, immediately brings back pain in the joints. He keeps the patient between blankets. As soon as the pain has gone, and the tongue is clean, he gives bitter tonics, omitting or diminishing the alkali if the urine is alkaline or neutral. If the tongue remains white after the pain has gone, he gives acids instead of alkalies with the bitter. Warm baths, he believes are useful and refreshing when the patient can be

moved without much pain. This treatment Dr. Farre has
adopted, with little variation, for many years, and is very
well satisfied with the result. The relief generally commences in forty-eight hours, often before. In some cases,
however, the rheumatism shows a disposition to return.
These, he treats, as Dr. NEVINS does, with quinine and iodide
of potassium, and, in most cachectic cases, gives quinine early
(as soon as the evacuations from the bowels are healthy,)
either with or without iodide of potassium. In the same
cases, too, he gives cod-liver oil. Iron he only uses when the
patients during convalescence are pallid. His treatment,
then, consists in calomel purges, bicarbonate or acetate of
potash, tincture of iodine or blisters; blankets; milk diet till
the pain subsides; then bitter tonics, with smaller doses of
potash, or with iodide of potassium, or with acids.

St. Thomas' Hospital.—The plan of treatment adopted by
Dr. PEACOCK, in cases of acute rheumatism, is chiefly the
alkaline and eliminative methods, giving full doses of the
bicarbonate of potash, with nitrate of potash, and, not unfrequently, iodide of potassium; and, in the latter cases, usually
combining the remedy with small doses of colchicum. Lat-
terly, he has employed blisters freely, in such cases as admitted
of their use; and provided several joints are affected, so that
four, or five, or six blisters can be applied at the same time, the
beneficial effect is most striking; the local symptoms are very
markedly and rapidly relieved, the constitutional disturbance
is lessened, and the disease cut short; so that cardiac symptoms are prevented or arrested, if in process of development.
He has not, except in very exceptional cases, relied wholly on
the local treatment; but has added it to the constitutional
measures which were previously in use; and the additional
benefit gained is often most striking. It is applicable especi-

ally to to the more intense cases of rheumatic fever; but is also very useful in those cases which are of such common occurrence, where the disease developes itself in persons previously most reduced in health and more particularly in persons who have previously had the disease, and often with cardiac complication. In such cases, if the disease be not rapidly arrested, the cardiac symptoms are almost sure to be aggravated; and the surest means of effecting that arrest he believes to be the use of eliminative treatment and free blistering. Such persons, also, should not be reduced if it can at all be avoided.

King's College Hospital.—The main points on which Dr. Johnson insists, are : that the patient should wear a large, loose, soft flannel dressing gown, instead of a cotton shirt; this should be changed at least every other day. If the pains be severe, he gives

53.1   R.   Pulveris opii,   gr.ss-j.
          Quiniæ sulphatis,   gr.ij.   M.
For one pill, ter die.

If the bowels be confined, a Seidlitz powder may be given every morning. He generally gives moderate doses of alkalies—one scruple or half a drachm of bicarbonate of potash with or without citric acid, every four or six hours. He is not satisfied that large doses of alkalies prevent cardiac complications; and he believes that they increase the tendency to rapid anæmia. In subacute cases, and in all cases where the skin does not act freely, he finds that hot air-baths are very useful. In cases of cardiac complications, especially pericarditis, with pain, he applies six leeches : then linseed poultices. He abstains from blisters and counter-irritation in the early stages of pericarditis. In cases of endocarditis, in order to lessen the tendency to deposit fibrin on the inflamed valves, he gives five-grain doses of sesquicarbonate of ammonia, with the alkaline mixture.

Middlesex Hospital.—Dr. GOODFELLOW, from a long experience, has eventually arrived at the conclusion that large and frequently repeated doses of alkalies, chiefly the nitrate of potash, in doses twenty grains at a time, with smaller doses of other alkalies, are more effectual in cutting short the attack and rendering the heart less liable to organic affection, than any other remedy. He, at the same time, applies cotton-wool to the præcordial region. If the joints be tense and painful, nitre poultices or wet compresses are applied; and, if they be less acutely affected, cotton-wool. He advocates flannel being worn to encourage perspiration. He strongly deprecates the practice of exposing the chest to the extent usually done, and percussing the præcordial region more than is absolutely necessary, as he believes that pericarditis may follow such a course, or at least, existing attacks may be increased in severity.

Westminster Hospital.—Dr. FINCHAM has employed the treatment by blisters for some time, and he is satisfied that, by this plan, the relief produced is very great, and the duration of the malady shortened. He is in the habit, however, as a rule, of combining with it alkalies in full doses ; *e. g.*

532.   R. Potassæ bicarbonatis,   ℨj.
       Potassæ nitratis,   ℨij.
       Liquoris ammoniæ acet ,
       Aquæ,   aa f.℥iij   M.
   A tablespoonful, in water, every four hours, with a full opiate at night, if the pain be severe.

He does not, however, think it advisable to continue the alkaline treatment for any lengthened period ; but to give quinine, in doses of two or three grains, every six hours, when the urgent symptoms begin to yield, especially if the sweating

12

is over profuse. He believes that, by giving quinine earlier than is generally the custom, convalescence is less tedious, and there is less chance of relapse. As regards cardiac complications, if pericarditis supervene, and there be sharp, catching pain, he applies a few leeches, followed by linseed poultices ; should the pain be slight or absent, he omits the leeches. In all cases he applies, subsequently, one or more blisters. As to medicines, he continues the alkalies, giving, at the same time, a grain of opium every four or six hours. Should endocarditis manifest itself, he contents himself with the alkaline treatment, as he cannot satisfy himself that local remedies have any effect.

In Dr. BASHAM's words typical cases are treated chiefly with salines ; either the nitrate of potash largely diluted, and given as a drink, with a little lemon juice ; or with the bicarbonate of potash and carbonate of ammonia, in a state of effervescence, with lemon juice. Dover's powder and nitre in equal portions, at bed time. If the alvine discharge at the outset be of the characteristic hard and offensive form, a brisk purge should be given.

Glasgow Royal Infirmary.—The treatment in rheumatic fever which Dr. GAIRDNER has usually followed, has been that by alkalies, and especially by acetate of potash, commonly aided by smaller doses of iodide of potassium, which last he began to employ systematically as part of the alkaline treatment since going to Glasgow, and much on the recommendation of Dr. RITCHIE, of that city, who was long in the habit of combining it with the acetate. The portion he usually employs is one drachm of iodide to one ounce of acetate in one pint of water, with any syrupy excipient that may be preferred to give flavor and take off the bitter saline taste. Lately he has tried the blister practice of Dr. HERBERT DAVIES, and, he thinks, with good success in some cases, cer-

tainly with manifest relief at the time. But he has not learned
to trust entirely to this treatment, and has used it only along
with the other. The joints are commonly wrapped in cotton
wadding, whatever the treatment in other respects. In a few
cases he has used considerable doses of quinine, in a few,
arsenic, in very many, opium, either as a principal or as an
accessory remedy, and often in pretty high doses.

Queen's Hospital, Birmingham.—At this hospital the number of cases treated is very large, and many are of great severity. The following treatment is that adopted by Dr.
FLEMING. The patient is placed between soft blankets, and
carefully protected from cold draughts. A meal is given
every four hours, consisting, during the fever, of milk and
strong beef-tea alternately. The diet is cautiously improved
during convalescence. One hour before each meal this draught
is administered.

533. R. Potassæ bicarbonatis,   gr. xxx.
       Aquæ,   f.ʒij.   M.
   For one dose; add one half an ounce of fresh lemon juice,
       and take the mixture during effervescence.

If there be high fever, from one to three minims of Fleming's tincture of aconite are added to each draught. If there
be much pain in the muscles, in place of aconite, from five to
ten minims of tincture of hemlock are added to each dose of
the alkaline. If, on the other hand, the periosteum be affected, from two to six grains of the iodide of potassium are
given. To relieve pain and secure sleep, Dr. Fleming orders,
at bed-time, a full draught of morphia and Indian hemp; as
a drink, potassa water or lemonade freely. If necessary, colocynth and hyoscyamus pill is given to relieve the bowels.
Cotton wadding is applied to the affected joints. Active and
repeated counter-irritation and poultices over the heart are

employed in cardiac inflammations. During convalescence, warm clothing, full diet, with quinine and iron. After considerable experience, Dr. Fleming has found that this treatment has furnished very good results, and that the number of those attacked with cardiac inflammation, *after* their admission into the hospital, is undoubtedly small. Placing the patient between blankets materially promotes perspiration, and prevents chills. In two recent cases, where this plan was followed, sudamina appeared over the entire surface. The contents of the vesicles were ascertained by Dr. SAWYER, the resident physician, to be alkaline, not acid.

In the *British Medical Journal*, for Jan. 16, 1869, p. 62, Dr. A. MYERS, Assistant Surgeon Coldstream Guards, remarks that of the hospitals quoted, in two only is the absence of sheets recommended, and in two the use of flannel apparel. He believes these points most essential to be attended to in aiding this disease to run a limited course with freedom from heart complication. He would, therefore, specially recommend :

1. That, in all cases, the patient should wear a flannel garment and be laid between blankets.

2. That, on the first evidence of pain in a joint, a thick layer of cotton wool should be smoothly wrapped round it and firmly bound with a flannel bandage. This, if the pressure be *equally* applied, gives immediate comfort to the patient, as well as keeps the affected joints at rest—a point so specially referred to, as of benefit, by Dr. WEBER.

3. That milk should be the chief article of diet, in most cases, during the early period of the disease ; its efficacy and appreciation by the patient being greatly increased when combined with soda or potass water in equal parts.

The *Lancet* gives the following treatment of acute rheumatism, as practised by Dr. SIBSON at

St. Mary's Hospital.—1st. Removal of pressure and tension of joints. 2d. An even and warm temperature. 3d. Removal or relief of pain. To accomplish the first of these ends, the patient lies in bed, and his joints are muffled in cotton-wool or flannel, a cradle being placed where the weight of the bed-clothes is painful. For the second, the patient wears a flannel dressing-gown, and the blankets touch the skin of the lower extremities, sheets being placed only over the upper part of the bed. For the third, the linimentum belladonnæ of the Ph. B. is applied to painful joints, and covered over with wadding. Occasionally, where the pain is very excessive, from an eighth to a quarter of a grain of morphia is injected subcutaneously. For the rest, he has now and then found it useful to apply a leech or two to a swollen joint, or to the cardiac region. In cases where there appears to be a gouty complication, Dr. Sibson employs a little iodide of potassium; but apart from this he does not give any potash to his patients. He finds the urine rarely containing acid after the first few days of treatment. As regards food, his experience and practice are not a little interesting. The patient is allowed from the first, roasted meat, rice pudding, and porter. This diet was not only ordered by the Doctor, but was consumed by the patient with very rare exceptions. Some patients confirmed this statement, and added also strong testimony to the immense relief derived from the application of belladonna in the way described.

GOUT.

JOHN HUGHES BENNETT, M. D., F. R. S. E., PROF. IN THE UNI-
VERSITY OF EDINBURGH.

534. R. Potassæ nitratis,   ℥ss.
          Aquæ,   f.℥vj.   M.
A tablespoonful every four hours.

Our author has employed this mixture in acute gout as well
as acute rheumatism with marked benefit in securing diaphoresis and relief of the pain.

535. R. Potassæ acetatis,   ℨijss.
          Spiritûs ætheris nitrosi,   f.ℨss.
          Tincturæ colchici,   f.ℨj.
          Aquæ camphoræ,   ad. f.℥viij.   M.
Two tablespoonfuls ter die.

536. R. Ammoniæ phosphatis,   ℨj.
          Tincturæ colchici,   f.ℨij.
          Aquæ,   f.℥vj.   M.
Two tablespoonfuls ter die.

In chronic gout with tophaceous deposits in the joints.

PROF. S. D. GROSS, PHILADELPHIA.

537. R. Vini colchici radicis,   f.ℨj.
          Morphiæ sulphatis,   gr.j.   M.
For one dose at bed time in gouty affections of the joints.

This treatment should be preceded by purgation or vene-
section, if indicated, and be followed in the morning by a gentle laxative. These doses are recommended by our author as
the most efficient, and as seldom disappointing the most san-
guine anticipations. When there is a full bounding pulse
indicating excessive arterial action, then the following will
come into play:

538. R . Tincturæ aconiti radicis,   f.ʒj.
          Morphiæ sulphatis,   gr.ij.
          Antimonii et potass. tart.   gr.j.
          Aquæ,   f.ʒjss.
          Syrupi zingiberis,   f.ʒss.   M.

A teaspoonful every three hours.

Veratrum viride may be substituted for the aconite in the same or double the dose. The action of these potent remedies should, of course, be carefully watched and kept within proper limits. Together with the above means the following should be employed to neutralize the acid state of the blood :

539. R. Potassæ bicarbonatis,   ʒj.
          Sodæ bicarbonatis,   ʒij.   M.

For six powders; one to be taken every six hours in a wine-glassful of water.

As a local application nothing will be found better than

540. R. Tincturæ opii,   f.ʒj.
          Linimenti saponis,   f.ʒij.   M.

To be rubbed in twice a day and constantly kept in contact with the affected joint by means of a piece of flannel covered with oiled silk A fly blister may be used if the disease manifests a disposition to linger.

LONDON HOSPITALS.

Charing Cross Hospital.—Dr. SALTER'S treatment of cases of acute gout does not differ in any essential particulars from the general management of such cases; and the results are such as, in his opinion, to entitle the treatment to be considered successful. It consists of the administration of certain remedies; the prescription of certain dietetic and other management; and the application to the part affected of a certain local treatment. What he generally orders is a mixture containing iodide of potassium, bicarbonate of potash, colchicum wine, and decoction of bark. He regards as groundless, in the great majority of cases, the fears that are so often expressed of the peculiarly lowering tendency of col-

chicum; at the same time recognizing the fact that cases are
sometimes met with which appear to be almost absolutely in-
tolerant of it, and others that bear it very ill. He thinks that
it should always be commenced very cautiously and tenta-
tively with those who have never taken it before. He is
equally incredulous of the opinion that has been expressed
by Dr. TODD and others, that colchicum tends to render gout
more inveterate and more apt to recur.

Dr. Salter thinks it very important, unless the case is
trifling, that the patient should be kept in bed, for the sake of
the perfect physical rest, for suspending all wear and tear,
and for getting some sleep by day in case the rest is much disturbed at night. He prescribes a light and simple diet—
faranacious foods made with milk, beef-tea, and fish. He
does not by any means consider stimulants a *sine qua non;* he
very often gives none at all; and in cases where the patient's condition absolutely requires it, he prefers claret, or
claret and potash water, to anything else. Unless the pain is
very severe and distressing by day, he does not give any sedative except at night, when he gives a sufficiently large dose
to command sleep, whatever that dose may be.

The local treatment of our author is all that is peculiar. It
consists in the application of the following lotion :

> 511. R.   Potassi iodidi,   ℥j.
>   Potassæ bicarbonatis,   ℥j.
>   Aquæ bullientis,   Oj.   M.
>   To this a little tincture of opium may be advantage-
>   ously added.

Doubled lint saturated with this lotion is applied to the
part affected, and covered with oil-silk ; to that is put a layer
of cotton wool, and the whole swathed in a flannel bandage.
The lint should be taken off from time to time, and re-dipped
in the lotion. The relief that the patients experience from

this application is very great. With or without this lotion, there are three other things on which Dr. Salter insists in the local treatment of a gouty joint—perfect physical rest, protection, and preventing the part affected being too dependent.

Middlesex Hospital.—In the treatment of acute gout, Dr. MURCHISON commences by clearing out the bowels with colocynth, blue pill and henbane, and then he relies mainly on alkalies and colchicum, the bicarbonate of potash and colchicum wine. With these he usually combines the nitrate of potash, and in private practice the patient is also instructed to drink lithia water. In rare cases there is irritability of the stomach; it may be necessary to subdue this by bismuth, magnesia, lime water and ice, with sinapisms to the epigastrium, before giving colchicum. The inflamed joints are covered with pledgets of lint moistened with laudanum, or with belladonna liniment and oiled silk, and the whole enveloped in cotton wool. Opiates are not given except in rare cases where the pain is protracted and severe, and not even then unless the bowels be well open, and the urine free from albumen. The patient's diet is restricted for the most part to milk and farinaceous articles.

J. SPENCE RAMSKILL, M. D., PHYSICIAN TO THE HOSPITAL FOR PARALYSIS AND EPILEPSY, LONDON, ETC.

Our author read at the meeting of the Harveian Society a paper on the therapeutic value of

OLIVE OIL.

The paper consisted of a history of two cases of gout, which he considered types of the kind of disease, and especially as to the stage of it in which the internal administration of olive oil was most useful. The first type was represented by a patient affected with comparatively acute attacks, reappearing with very short intervals, and making little or no way toward convalescence. Bark, quinine, iron had frequently failed to prevent a reappearance of the disease. Cod-liver oil was rarely borne at all. In such cases, olive oil, given when the patient lapsed in the interval, had answered all the requirements of the case in Dr. RAMSKILL's hands. Nutrition began to improve, and no more relapses occurred. The second case was a type, also, of a class of cases, where all acute symptoms having long subsided, vague and uneasy pain remained in all the joints—associated only with stiffness or difficulty and pain on movement. The general health, meanwhile, slowly deteriorated, with much general wasting; and no impression could be made on the system by the usual tonics. Here the use of olive oil was more quickly beneficial; but it often seemed to act as a hæmatogen. In true rheumatoid arthritis, the use of the oil was, perhaps, more beneficial than most ordinary remedies; but Dr. Ramskill could make no assertion as to the favorable action of any single remedy on this disease. The dose of olive oil should not exceed a teaspoonful at the commencement; it should be gradually increased until a laxative effect announced the attainment of such a dose as exceeded the absorbent power of the stomach and intestines. Any vehicle, containing a few drops of sulphuric ether, would then help to assimilate the oil and prevent diarrhœa. It was important to obtain perfectly fresh and new oil, to insure absence of rancidity, and consequent eructations and disorder of the stomach. Dr. Ramskill considered the remedy as a combination of food and physic; but still one unattainable by ordinary food and medicine. It was important to begin its administration when the patient was free from acute attacks, or, at least, from fever. The passage of pale urine, or of greenish yellow urine, that suggesting oxaluria, was an indication for its use, especially if accompanied by hypochondriasis, general *malaise*, and weariness and aching of joints. Dr. Ramskill said he had found great benefit from the use of olive oil at the Hospital for Paralysis and Epilepsy, especially in cases of lead-poisoning, after the acute symptoms, such as colic, had subsided; always in the malnutrition accompanying paralysis of the extensors of the hands; also in Cruveilhier's atrophy; and in epilepsy, associated with great cachexia. In all these conditions, supposing cod-liver oil disagreeing, and, therefore, inadmissible.

St. George's Hospital.—For the purpose of clinical instruction, Dr. FULLER divides cases of acute gout into two classes, namely : Cases in which the excretory organs are originally sound and functionally active—cases in which the attack of gout is due principally to excess and indiscretions of diet; and 2. Cases in which the excretory organs are in some way disordered, and fail in performing their eliminatory functions—cases in which the patient is not necessarily guilty of indiscretions of diet, but in which the liver and kidneys fail in their action, either as the result of functional disorder, or of organic change in their structure.

The first class of cases correspond with those which pass under the name of asthenic gout; the tongue is usually furred; the urine loaded, and the bowels are commonly torpid. In these cases until the acute symptoms have subsided, Dr. Fuller restricts the diet to liquids, administers a saline draught containing sulphate and carbonate of magnesia, and a few drops of colchicum wine, occasionally gives an aperient pill containing calomel, aconite and opium, and wraps the joints in finely carded wool, or in flannels steeped in a solution of soda and laudanum. As the acute symptoms subside, a more generous diet is permitted, and some light, bitter tonic, such as tincture of gentian or calumba, is added to the mixture.

The second class of cases have more affinity with what is termed atonic gout: the tongue is often clean and the urine clear—sometimes of low specific gravity,—and the bowels are regular. In these cases Dr. Fuller does not restrict the diet to the same degree; he allows a little meat without vegetables, and also, if desired, a glass of sherry or a little spirits and water. He acts freely on the skin by means of the hot-air bath; administers an aperient in the morning containing taraxacum and sulphate of magnesia, and during the day he gives a warm stomachic draught containing ammonia, and a few grains of soda in a light bitter infusion. Occasionally a dinner pill is prescribed containing rhubarb and a grain of colchicum; and in some instances, characterized by pale clear urine, a draught containing quinine, the mineral acids, and taraxacum, is substituted for the mixture just referred to. In these cases, as soon as the acute symptoms have subsided, a drachm of the syrup of phosphate of iron is given each morning before breakfast.

Westminster Hospital.—Dr. RADCLIFFE thinks that, during the last 20 years, there has been a great change in the character of the cases of gout which fall under the physician's notice. The acute gout of old, he believes, is now rarely met with. It is much more common to meet with the subacute form—the form, that is, which is more nearly allied to rheumatic gout. Dr. Radcliffe does not employ colchicum. In a case of gout where some part of the foot is involved, he raises the limb to a height above that of the pelvis, gives diluents, iodide of potassium, alkalies, and no colchicum. Nor does he give purgatives. He diminishes the allowance of port wine and beer.

ANÆMIA.

PROF. JOHN B. BIDDLE, PHILADELPHIA.

512. R. Quiniæ sulphatis, gr. ij
      Ferri sulphatis, gr. j.
      Strychniæ, gr. 1-60. M.

For one pill ter die.

An excellent tonic combination frequently prescribed by our author.

THOMAS K. CHAMBERS, M. D.

TONIC BATH.

543. R. Acidi muriatici, f.ℨj.-ij.
      Aquæ calidæ, C.xxx. M.

In a wooden bath, the patient to remain in it for from ten to twenty minutes.

Dr. CHAMBERS has found this bath to confer muscular strength, even when employed alone. When used in cases of anæmia while iron is being given internally it gives an impetus to the improvement of the patient; more iron is taken up, and the blackening of the fæces ceases.

J. M. DA COSTA, M. D., OF PHILADELPHIA.

TONIC INHALATION

544. R. Ferri lactatis, gr. j.-ij.
      Aquæ destillatæ, f.ℨj. M.

For one inhalation. To be administered (by means of any form of steam atomizer throwing a fine spray) two or three times a day. Useful in anæmia, when iron given by the stomach is not assimilated. Also in cases of gastric ulcers, when the constitutional effect of this agent is called for, while it is desirable to spare the stomach.

545.   ℞ ·   Ferri pyrophosphatis,   ℥j.
       Aquæ bullientis,   f.℥ss.   M.
         And add,
       Extracti gentianæ fluidi,   f.℥ss.
       Curaçoæ,   f.℥iss.
       Vini,   q. s. ad fiat, f.℥iv.   M.
    Teaspoonful ter die.

This preparation, known as elixir gentianæ ferratæ, is much used and highly esteemed in this city.

PROF. EASTON, UNIVERSITY OF GLASGOW.

The following formula of our author has become very popular in England, since its first publication in Aitken's Practice of Medicine, as a general tonic in anæmia and cachexia. It is known as the

Syrupus Ferri Quiniæ et Strychniæ Phosphatum.

546.   ℞ ·   Ferri sulphatis,   ℥v.
       Sodæ sulphatis,   ℥vj· vel ℥j.
       Quiniæ sulphatis,   gr. ccxij.
       Acidi sulphurici diluti,   q. s.
       Aquæ ammoniæ,   q. s.
       Strychniæ,   gr. vj.
       Sacchari albi,   ℥xiv.
       Acidi phosphorici diluti.   f.℥xiv.

    Dissolve the sulphate of iron in one oz. boiling water and the phosphate of soda in two oz. boiling water. Mix the solution and wash the precipitated phosphate of iron till the washings are tasteless. With sufficient dilute sulphuric acid dissolve the sulphate of quinia in two oz. water. Precipitate the quinia with ammonia water and carefully wash it. Dissolve the phosphate of iron and quinia thus obtained, as also the strychnia in the diluted phosphoric acid; then add the sugar and dissolve the whole, and mix without heat.

The above syrup contains about one grain phosphate of iron, one grain phosphate of quinia and one thirty-second of a grain of phosphate of strychnia in each drachm. *The dose* might, therefore, be a teaspoonful three times a day.

The amount of phosphate of quinia might be increased according to circumstances; and if eight grains of strychnia were employed in place of six, as in the above. the phosphate of strychnia would be in the proportion of one twenty-fourth of a grain in every fluid drachm of the syrup. A much larger dose should scarcely be ventured upon.

PROF. S. D. GROSS.

547. R. Tr. ferri chloridi,   f.ʒj.
          Quiniæ sulphatis,   gr. xx.   M.

Sig. Twenty drops ter die, in sweetened water, through a tube.

He prefers the tincture of the chloride to all the other preparations of iron.

Prof. CHAS. D. MEIGS considered reduced iron (ferrum redactum) to be the most efficient of the chalybeates, in two-grain doses three times a day, after each meal, on a full stomach.

DR. JOHN FORSYTH MEIGS, PHILA.

518. R. Ferri et quiniæ citratis,   ℈. iv.
          Extracti gentianæ fluidi,
          Spts. lavand. compositi, aa f.ʒiij.
          Alcohol,   f.ʒvj.
          Aquæ,   f.ʒivss.   M.

A tablespoonful ter die.

PROF. ELLERSLIE WALLACE, PHILA.

519. R. Ferri pyrophosphatis,   ʒij.
          Curaçoæ,   f.ʒss.
          Aquæ,   f.ʒijss.   M.

Sig.—Teaspoonful four times a day.

CHLOROSIS.

WILLIAM AITKEN, M. D., EDINBURGH.

The food of chlorotic patients must be regulated so as neither to be too stimulating nor disgustingly mild and so as to secure a variety. Half an hour before each of the meals order

550. R.   Quiniæ sulphatis,   gr.j.
         Pulveris capsici,   gr.ij.   M.
         For one pill.

The various preparations of iron are useful as stimulants to digestion, etc. The syrup of the phosphates (F. 546) may be given. The eliminative action of the colon is to be promoted by

551. R.   Pilulæ aloes et myrrhæ,   gr.iv.
         For one pill, every night at bed-time.

Simple bitter tonics are useful adjuncts to the chalybeate treatment, such as *gentian, columba,* and the preparations of *cinchona.* They aid feeble digestion.*

PROF. T. GAILLARD THOMAS, M. D., NEW YORK.

Our author, who considers this disease as a *neurosis* of the ganglionic system of nerves, states that the treatment should not consist in fruitless attempts to overcome one or even two of the results of the disease, amenorrhœa and anæmia, but in a systematic effort to accomplish these three ends:

1st. To remove the cause of the disorder.

2d. To cure the neurosis itself.

3d. To repair the damage which it has effected in the system.

---

*Science and Practice of Medicine, 2d Am. Ed., p. 100.

Search should always be made for the cause ; when it cannot be found or removed, much good may be accomplished by sending the patient away from home. A sea-voyage and visit to a foreign country are often efficacious. Well regulated exercise with open air is of importance, but inferior to cheerful, congenial, and new society.

In the meantime, nervous tonics of medicinal kind should be freely given. Preparations of arsenic, strychnia, and quinine are the best. General electrization is often very beneficial.

As anæmia is usually a complication, chalybeates are indicated. The saccharated carbonate, reduced iron and the bitter wine of iron are among the best. A very excellent combination is offered by the following prescription :

552. R. Liquoris potassæ arsenitis, f.ʒij.
Tincturæ nucis vomicæ, f.ʒss.
Vini ferri amari, f.ʒvijss. M.
A dessertspoonful, in a claret-glassful of water, just after each meal.

The *diet* should be extremely nutritious, *i. e.*, meat, eggs, animal broths, and vegetables, with wine, whiskey, or malt liquors, if these are indicated by great exhaustion.*

VII. DISEASES OF THE SKIN.

ERYTHEMA.

J. M. DA COSTA, M. D., PHILADELPHIA.

553. R. Unguenti picis,
Ung. hydrargyri oxidi
rubri. aa. ʒss. M.
To be applied morning and evening in *chronic erythema.*
Internally, *Donovan's solution,* gtt. x, ter die.

---

*Practical Treatise on the Diseases of Women. 2d ed., p. 631.

In *acute erythema*, a useful sedative ointment is

   554. R.   Liq. plumbi subacetatis,
               Glycerinæ,   aa. f.ʒj.
               Cerati simplicis,   ʒvj.   M.

or,

   555. R.   Cerati plumbi subacetatis, ʒvj.
               Glycerinæ,   f.ʒij.   M.

TILBURY FOX, M. D., LONDON, ETC.

In the local erythemata we must first remove all irritants, pay especial attention to cleanliness and merely apply soothing agencies, e. g., to prevent dryness or friction, etc., zinc ointment (567), or glycerine and rose water; linimentum aquæ calcis; fine starch or lycopodium powders; avoid poultices, and give aperients internally.

In mild cases of *intertrigo* the same plan of treatment is adopted. In troublesome cases, with sour acrid discharge, alteratives with chlorate of potash, in the first instance, internally, are of service, together with a nutritious milk diet. Then, locally, zinc ointment, starch powder, or

BISMUTH LOTION.

   556. R.   Bismuthi subnitratis,   ʒij.
               Hydrarg. chlor. corrosivi, gr.x.
               Spiritûs camphoræ,   f.ʒss.
               Aquæ,   Oj.   M.
     Apply, diluted with one, to two, or three parts of water.

Or,

CALAMINE LOTION.

   557. R.   Zinci oxidi,   ʒij.
               Zinci carbon. precipitatæ, ʒij.
               Glycerinæ.   f.ʒij.
               Aquæ rosæ,   f.ʒviij.   M.

Or, lastly, if the case be chronic, a weak solution of nitrate of silver may be used. Syrup of iodide of iron and cod-liver oil are also called for.

FOR CHILBLAINS.

558. R. Olei terebinthinæ,
          Tincturæ aconiti,
          vel
          Tincturæ belladonnæ,
          Linimenti saponis,   aa f.ʒj.   M.

This, our author says, is the best treatment, together with
iron, quinine, and cod-liver oil.* See also p. 196.

DR. FELIX VON NIEMEYER, PROFESSOR UNIVERSITY OF
TUBINGEN.

Erythema resulting from local irritation soon disappears
spontaneously, the cause being removed. Applications of
cold water or lead water relieve the burning pain if severe.

In erythema intertrigo, in order to prevent the friction of the
opposing surfaces, they ought to be sprinkled with a fine
powder. Oxide of zinc and lycopodium are the best, (see F.
559). A pledget of charpie, smeared with zinc ointment,
(F. 567), way be inserted between the surfaces.

Erythema arising from the pressure against the bed may be
relieved by air pads of india rubber. When it results from
the contact of acrid secretion, the skin is to be protected by
a coating of lip salve or other grease.

In erythema nodosum proper attention must be paid to the
fever and to the strength of the patient. Compresses wet
with cold water, or with lead water, should be applied to the
nodules if painful.†

PROF. J. LEWIS SMITH, NEW YORK.

559. R. Pulveris zinci oxidi,
          Lycopodii,   aa. ʒj.   M.
   To be dusted occasionally over the inflamed surface in the
   erythema intertrigo of infancy, when the inflammation is
   severe and accompanied by moisture.

*Skin Diseases, London, 1860, p. 75.
†Text-Book of Practical Medicine. Am. ed., vol. ii, p. 109.

In slight cases of this affection, due to friction of opposing surfaces of the skin, or to the irritation of certain discharges, if not accompanied by moisture and destruction of the epidermis, dusting the surface thickly with *powdered starch*, so as to prevent attrition, will be all the treatment required. The disease may also be satisfactorily treated in most cases by the following wash:

560.   R.   Cupri sulphatis,   gr.ij-iv.
             Aquæ rosæ,   f.ʒij.   M.
       To be kept constantly applied by means of linen saturated
          with it and pressed between the inflamed surfaces.

When this disease is caused by frequent acid stools, remedies which cure the diarrhœal affection also cure the erythema.[*]

BALMANNO SQUIRE, M. B., F. L. S., ETC.

TREATMENT OF CHILBLAINS.

Measures must be adopted to increase the activity of the general circulation by a generous and stimulating diet, active exercise, frictions of the skin with hair gloves, etc., and at the same time, activity of the circulation in the affected part should be specially promoted by the use of stimulating applications, such as soap liniment or camphor cerate. If the chilblain be "broken," resin ointment will be a suitable dressing; poultices are to be avoided, if possible.[†]   (See F. 558, 568.)

THOMAS HAWKES TANNER, M. D., F. L. S., LONDON, ETC.

In the treatment of erythema any derangement of the digestive, urinary or uterine functions, which may exist, must be removed. The administration of a mild saline aperient, warm water, or vapor baths, light diet, tonics, especially qui-

---

[*] Diseases of Infancy and Childhood, 1869. p. 559.
[†] Manual of the Diseases of the Skin. Lond., 1869, p. 9.

nine, with compound tincture of bark or the mineral acids, are sufficient for the cure of most cases. In some varieties a local application may be required. Then the *liquor plumbi subacetatis* can be used.

In *erythema nodosum* use

561. ℞. Veratriæ,   gr.viij.
    Adipis,   ʒj. .
    Olei olivæ,   f.ʒss.   M.
   Rub the veratria and oil together; then mix them
   thoroughly with the lard.*

This ointment will relieve the tenderness while quinine is being administered to effect a cure.

ERASMUS WILSON, F. R. S., ETC., LONDON.

562. ℞. Olei juniperi pyrolignici,
    Alcoholis,   aa f.ʒj.
    Saponis mollis,   ʒij.   M.
   As a wash in *erythema intertrigo.* The parts are afterwards
   to be dressed with

563. ℞. Ung. zinci oxidi benzoati,   ʒij.
    Spiritûs camphoræ,   f.ʒij.   M.

564. ℞. Ung. zinci oxidi benzoati,   ʒij.
    Liq. plumbi subacetatis,   f.ʒij.   M.

This is a soothing application for erythema of the *vulva and anus.* Or the following may be used:

565. ℞. Pulveris plumbi acetatis,   gr.xij.
    Unguenti benzoati,   ʒj.   M.
   Mix thoroughly.

Over these an evaporating lotion may be employed, if requisite. Nitrate of mercury ointment, more or less diluted, may replace the above after the acute stages have passed.

The formulæ for the simple benzoated ointment and for the benzoated ointment of the oxide of zinc, recommended by Prof. Wilson in the above recipes, are as follows:

*Practice of Medicine, Am. Ed. p. 657.

UNGUENTUM BENZOATUM.

566. ℞. Adipis purificatis,   ℥vj.
Gummi benzoini pulveris,   ℥j.   M.

Rub together; afterwards melt with gentle heat, for
twenty-four hours, in a closed vessel, and strain
through linen.

UNGUENTUM ZINCI OXIDI BENZOATUM.

567. ℞. Adipis preparati,   ℥vj.
Gummi bezoini pulveris,   ℥j.   M.

Melt with gentle heat for 24 hours in a closed vessel;
then strain through linen and add

     Zinci oxidi purificati,   ℥j.
Melt well and press through linen.

568. ℞. Spiritûs terebinthinæ,
Acidi acetici diluti,   aa f.℥j.   M.
Add the contents of one egg and shake well together.

A stimulating liniment useful in the erythematous state of
chilblains.*

ROSEOLA.

PROF. J. LEWIS SMITH, NEW YORK.

569. ℞. Liq. ammoniæ acetatis,
Misturæ camphoræ,   aa. f.℥iv.   M.

As a lotion in *roseola infantalis*, when there is itching or
tingling of the surface. It is to be used lukewarm.

Or,

570. ℞. Acidi hydrocyanici dil.,   f.℥j.
Emul. amygdalæ amaræ,   Oj.   M.
To be used warm.

Cold applications, which would repel the eruption, should
be avoided. In all cases the state of the health should be
inquired into, and any deviation from the normal condition

*Treatise on the Diseases of Infancy and Childhood. 1869, p. 560.

corrected. No further constitutional treatment will be required.*

ERASMUS WILSON, F. R. S., LONDON, ETC.

Gentle laxatives, effervescent salines, light bitters with the mineral acids, small doses of quinine with sulphuric acid, mild chalybeates, constitute the pharmacopœia of roseola, both in its idiopathic and chronic form. Locally, if much irritation be present, the skin may be washed with juniper tar soap or carbolic acid soap and tepid water, or sponged with hot water, or with

571. R. Ammoniæ carbonatis, ʒj.
Aquæ, Oss. M.
To be used tepid.

Or, the following:

572. R. Acidi hydrocyanici dil., f.ʒj.
Emul. amygdalæ amaræ, f.ʒvj. M.

Where any fear of repercussion of the exanthem exists, benzoated oxide of zinc ointment (F. 567) should be gently rubbed into the skin.

The *diet* should be antiphlogistic; toast-water and barley-water, with or without chlorate of potash or lemon juice, for drinks; with milk farinacious puddings, broths, eggs, fish, poultry; returning by degrees to the accustomed diet.†

URTICARIA.

WILLIAM AITKEN, M. D., EDIN.

In the treatment of nettle-rash, emetics and purgatives are to be employed in the first instance; afterwards faulty digestion is to be corrected. The surface of the eruption may be dusted with flour, or the following lotion may be used:

*Diseases of the Skin. 7th Am. Ed., pp. 221, 222.
†Diseases of the Skin. 7th Am. Ed., p. 257

573. ℞. Ammoniæ carbonatis,   ʒj.
          Plumbi acetatis,   ʒij.
          Aquæ rosæ,   f.ʒviij.   M.*

Our author finds urticaria one of the most difficult and unsatisfactory of all diseases to cure. The acute are more satisfactory to treat than the chronic cases.

The following is a resumé of what appears best to be done:

Urticaria Febrilis.—In simple cases use saline aperients, milk diet, no stimulants, alkalies largely diluted, alkaline baths, as

574. ℞. Sodæ carbonatis,   ʒviij.
      In an ordinary hip bath, twice a day.

The following lotion is serviceable:

575. ℞. Hydrarg. chlor. corrosivi, gr.jss.
          Chloroformi,   ♏xx.
          Glycerinæ,   f.ʒij.
          Aquæ rosæ,   f.ʒvj.   M.

So also is

576. ℞. Potassii cyanidi,   gr. vj.
          Cocci cacti,   gr. j.
          Unguenti aquæ rosæ, ·   ʒj.   M.

ALKALINE BATH.

577. ℞. Sodæ carbonatis,   ʒiv-viij.
          Potassæ carbonatis,   ʒiij-vj.
          Sodæ biboratis,   ʒij.
          Aquæ,   cong. xxx.   M.

If the patient be gouty, colchicum should be given with salines; when fever runs high give acetate of potash, tincture of digitalis, with even tartar emetic. The tincture of veratrum viride is then useful.

*Science and Practice of Medicine. Am. ed. vol. ij. p. 965.

Urticaria ab ingestis.—An emetic (zinc or ipecacuanha), a saline purge, and subsequently a mixture of carbonate of ammonia, prussic acid and infusion of cascarilla, say :

578. R.   Ammoniæ carbonatis,   gr.xxv.
           Acidi hydrocyanici,   ♏xij.
           Infusi cascarillæ,   f.℥vj.   M.
     A tablespoonful every four hours.

Chronic Urticaria.—The treatment is most tiresome and difficult. One has to analyze carefully every function of the patient. If there be mental disturbance, change of scene does good. Pyrosis, atonic dyspepsia, deficiency of bile, inaction of the liver, non-excretion of urea, uterine disorder must be treated upon general principles. Generally speaking, it is possible to discover some one thing, which, taken internally, evokes the urtication; it may be beer, or condiments of some kind. Where it appears that the functions of the body generally are properly performed, bromide of ammonium, or if the disease be periodic, quinine is useful ; aconite is another remedy ; arsenic is much vaunted, but our author is not very partial to it.*

THOMAS HUNT, F. R. C. S., ETC., LONDON.

579. R.   Liq. potassæ arsenitis,   f ℨjss.
           Liquoris potassæ,   f.℥ij.
           Tincturæ cardamomi,   f.℥ijss.   M.
     A teaspoonful ter die in chronic urticaria.†

THOMAS HAWKES TANNER, M. D., F. L. S., ETC., LONDON.

580. R.   Bismuthi subnitratis,
           Magnesiæ carbonatis,   aa. gr.x.   M.
     For one powder, to be taken in half a bottle of soda water, ter die.

In *chronic urticaria* iron will often effect a cure. The following recipe may be used :

---

*Skin Diseases. Lond. 1869, p. 88.
†Guide to the Treatment of Diseases of the Skin. Lond., p. 98.

581. ℞. Spts. ammoniæ aromatici, f.ʒss.
      Ferri et ammoniæ cit., Ɉij.
      Infusi quassiæ, f.ʒvjss.
      Glycerinæ, f.ʒj.   M.

Two tablespoonfuls ter die.

582. ℞. Tincturæ ferri chloridi, f.ʒjss.
      Acidi muriatici diluti, f ʒij.
      Tincturæ hyoscyami, f.ʒiij.
      Aquæ camphoræ, q. s. ad. f.ʒvj   M.

A tablespoonful ter die.

583. ℞. Ferri et ammoniæ cit., ʒj.
      Spts. ammoniæ aromat., f.ʒss.
      Potassæ bicarbonatis, ʒij.
      Infusi calumbæ, q. s. ad. f.ʒvj.   M.

Two tablespoonfuls to be taken twice a day, with one tablespoonful of lemon juice.

In obstinate cases, where there are no symptoms of gastro-
intestinal irritation, small doses of arsenic may required, as,

584. ℞. Liq. potassæ arsenitis, ♏Ɉ.
      Tincturæ lupuli, f.ʒj.
      Infusi qusssiæ, f.ʒiij.   M.

A dessertspoonful three times a day, directly after meals.
The dose should be diminished so soon as the tongue gets
thoroughly coated with a silvery-looking fur, or the conjunctivæ become irritable, or diahrrhœa sets in, or gastric pain is complained of.

The irritation can be relieved by sponging with leadwater,
or with equal parts of vinegar and water, or by,

585. ℞. Hydrarg. chlor. corros. gr.viij.
      Aquæ destillatæ, f.ʒviij.   M.

To be frequently applied.

ERASMUS WILSON, F. R. S., LONDON, ETC.

In chronic urticaria, the deranged functions are to be restored. The administration of the mineral acids with a bitter
is serviceable. Very chronic cases require arsenic. The following may be used:

586. ℞. Liquoris arsenici chloridi, f.℥ss.
Acidi muriatici diluti,
Aquæ flor. aurantii,   aa f℥ij.
Syrupi simplicis,   f.℥iij.   M.

A tablespoonful to be taken alone or in water, *with the meals,*
three times a day.

The local treatment consists in the use of remedies for the
purpose of relieving the itching, tingling, and smarting. For
this purpose employ sponging with hot water; ablution with
the juniper tar or carbolic acid soap; sponging with the juniper tar lotion (F. 562); frictions with

UNGUENTUM PICIS JUNIPERI.

587. ℞. Olei juniperi pyrolignici, f ℥j.
Adipis purificatæ,   ℥ij.
Sevi ovilli purificati,   ℥vj.   M.

Melt with gentle heat and make an ointment.

This is an elegant preparation. It may be used of the above
strength or diluted. Or, the lotion of emulsion of bitter
almonds with hydrocyanic acid (F. 670); or the

LOTIO HYDRARGYRI BICHLORIDI.

588. ℞. Amygd. amarum,   no. xx.
Aquæ destillatæ,   f.℥vj.

Contuse and mix together, then strain and add

Hydrarg. chloridi cor.,   gr.xvj.
Spiritûs vini rectificati,   f.℥ij.   M.

Or the

LOTIO ACIDI CARBOLICI.

589. ℞. Acidi carbolici fluidi,   f.℥ss.-j.
Glycerinæ,   f.℥ss.
Aquæ destillatæ,   f.℥vijss.   M.

Or, sponging with hot vinegar, with a lotion of carbonate
of ammonia, a lotion of aconite, and liniments of opodeldoc
and chloroform or laudanum. When one application fails the
other must be tried. The tepid bath affords almost instan-
aneous relief.*

*Diseases of the Skin. 7th Am. ed., p. 245.

PAPULAR DISEASES.

LICHEN.

TILBURY FOX, M. D., ETC., LONDON.

The early stages of lichen, when accompanied by febrile symptoms, are to be treated upon general principles. Salines and aperients are proper, together with tepid

EMOLLIENT BATHS.

590.	R.	Bran,	lb.2 to lb.6.	
		Water,	gal s. 30.	M.

Or,

591.	R.	Gelatine,	lb.1 to lb.3.	
		Water,	gal's. 30.	M.

Or,

592.	R.	Linseed,	lb.1.	
		Water,	gal's. 30.	M.

In lichen agrius poulticing, rest, and

593.	R.	Liq. plumbi subacetatis,	f ʒj–ij.	
		Infusi altheæ,	Oj.	M.

Use as a lotion.

Or, employ an ointment containing the watery extract of opium and lead. To allay itching at this stage, besides the baths, the following ointment is useful:

594.	R.	Potassii cyanidi,	gr.iij.	
		Adipis,	ʒj.	M.

Or direct,

595.	R.	Zinci oxidi,			
		Sodæ biboratis,	aa.	ʒj.	
		Camphoræ,		gr.x.	
		Adipis,		ʒj.	M.

Or,

596. ℞. Acidi hydrocyanici dil., f.ℨij.
Sodæ biboratis, ℨss.
Aquæ rosæ, f.ℨviij. M.

Or,

597. ℞. Hydrarg. chloridi corros., gr.j
Acidi hydrocyanici dil., f.ℨj.
Emul. amygdalæ amaræ, f.ℨvj. M.

Then, when the disease has passed the acute stage, the patient must be treated according to his constitutional bias. In a goodly number of cases it will be noted that he or she is overworked, worried, not taking sufficient food or rest, is annoyed by dyspepsia, and is looking thin and anxious. In such cases, a change from depressing and over work, the correction of acid or atonic dyspepsia, mild aperients and a course of mineral acids and bitters, will speedily be effectual, the local treatment consisting in the use of mild astringents, such as,

598. ℞. Zinci oxidi, ℨij.
Glycerinæ, f ℨij.
Liq. plumbi subacetatis, f ℨjss.
Aquæ calcis, f.ℨvj.–viij. M.

Or,

599. ℞. Acidi nitrici diluti, f.ℨjss.–ij.
Aquæ, f ℨvj. M.

In other cases where the urine is loaded and the skin generally is discolored and harsh, alkalies are of service, and may be given with ammonia and bitters, together with alkaline baths and borax lotions.

In other cases, it is apparently impossible to say that anything beyond general debility exists; under such circumstances, arsenic is to be employed.

600. ℞. Sodæ arseniatis, gr.j.–ij.
Aquæ destillatæ, f.ℨviij. M.

One teaspoonful at first, daily; then two in conjunction with, alternately, alkaline and vapor baths.

In *lichen circumscriptus*, an alkaline course is beneficial; and if there be any tendency to rheumatism, bromide of potassium may be given in addition. In this variety of lichen the following ointments are serviceable :

601.	℞.	Ung. hygrarg. nitratis,	ʒij.	
		Adipis,	ʒvj.	M.

602.	℞.	Ung. hydrarg. ammoniati,	ʒj.	
		Adipis,	ʒvij.	M.

In *lichen agrius* maceration with glycerine, or the following is useful :

603.	℞.	Sodæ biboratis,	ʒj–ij.	
		Glycerinæ,	f.ʒj.	
		Adipis,	ʒj.	M.

Or, paint with

604.	℞.	Argenti nitratis,	gr. ij–x.	
		Aquæ,	f.ʒj.	M.

When the disease is very chronic and there is much thickening of the skin in general, and in *lichen pilaris*, a course of bicyanide of mercury is necessary.

605.	℞.	Hydrargyri bicyanidi,	gr. j.	
		T᷉ cinchonæ comp.,	f.ʒiv.	M.
	A dessertspoonful ter die.			

This will cause an absorption of the plastic material poured out into the derma; and local stimulation to the skin with sulphur vapor baths may then be employed.

No one plan can be laid down for lichen. Each patient must be treated according to his individual peculiarities. The tendency should be in the early stage to use alkalies, and in the latter stage arsenic. The too free and early use of stimulants to the skin should be avoided—emollient and alkaline baths being most fitting for recent cases. In all, stimulants are to be dispensed with entirely, if possible, and the food is to be unstimulating.

In *lichen urticatus* the presence of scabies must be very
closely looked after. Our author has generally succeeded in
curing this obstinate form of lichen by insuring perfect cleanliness in the way of linen, giving diuretics, with occasional
doses of calomel, cod-liver oil and baths containing sulphuret
of potassium. The following recipes are useful in various
forms of lichen :

606. R. Plumbi carbonatis, ℈ss.
Chloroformi, ℳ.iv.
Unguenti aquæ rosæ. ℥j. M.
To allay itching.

Or, for the same purpose,

607. R. Chloroformi, ℳ.viij.
Potassii cyanidi, gr.iv.
Glycerinæ, f.℥j.
Cerati simplicis, ℈vij. M.

Or, again, to fulfill the same indication,

608. R. Hydrarg. chlo. corrosivi, gr.jss.
Chloroformi, ℳxx.
Glycerinæ, f.℥ij.
Aquæ rosæ, f.℥vj. M.

609. R. Magnesiæ sulphatis, ℥jj.
Magnesiæ carbonatis, ℈j.
Tincturæ colchici, f℈ij.
Olei menthæ piperitæ, ℳiij.
Aquæ, q. s. ad. f.℥iv. M.
A desertspoonful or tablespoonful in acute forms of the dis-
ease in loaded habits.

610. R. Strychniæ, gr.¼.
Acidi phosphorici diluti, f.℈j.
Tinct. aurantii corticis, f.℈ij.
Infusi caryophylli, f.℥iv. M.
A tablespoonful, ter die.*

*Skin Diseases. Lond. 1869, p. 133.

THOMAS HUNT, F. R. C. S., ETC.

In cases of lichen in which the disease shows no inclination to yield in a week or two, a steady, well regulated course of arsenic will eventually succeed. But it is useless to give this remedy for a few days or even weeks by way of *trial*. It must not be tried but *trusted* if it is to do its work, and several months perseverance may be requisite.

MANNER OF ADMINISTERING ARSENIC.

There are few medicines less likely to do harm than arsenic when administered in the manner about to be described. *Its curative powers seem to reside alone in doses too small to be mischievous.* It is impossible to push it. But a patient administration of small doses under favorable circumstances, for weeks, months, or years together, will be found to exercise an almost omnipotent influence over the cutaneous diseases to which it is adapted.

The numerous failures of arsenic may be traced to one or more of the following sources : 1. The syphilitic character of the cutaneous disease ; mercury is then wanted, arsenic having no influence whatever. 2. The administration of arsenic during the inflammatory or febrile stage of cutaneous disease, under which circumstances it rarely fails to increase the inflammation and never does any good. 3. Its administration on an empty stomach, thus exciting gastric irritation. 4. Too large doses and too long intervals between the doses. 5. The serious error of directing *gradually increasing doses.* The proper method is to increase the dose one-fifth, once or twice a month, if after a fortnight or three weeks it produces no sensible effect whatever. So soon as it begins to assert itself the full dose is arrived at and it should be continued without further increase. Five minims of Fowler's solution

ter die is sufficient to begin with, and this may be reduced as
occasion may require. It should be mixed with a little
water, or with the beverage drank with or after meals. Children above 5 years old will bear nearly as large a dose as adults.

A full dose being first administered at regular intervals,
in a few days (or possibly weeks) a pricking sensation is felt
in the tarsi, and the conjunctivæ become slightly inflamed.
*At this crisis the disease is brought under arrest and generally
from this period appears to be shorn of its strength.* The dose
may now be reduced, and in some cases a very small dose taken
with exact regularity will suffice to keep the eyelids slightly
tender and the skin healing, until at length, even the disposition to disease appears to die away under the influence of
the medicine. The patient should be examined at first once a
week. The medicine must not be entirely abandoned *until
weeks or months after all disposition to morbid action appears to
have subsided.* The arsenical course should be protracted, in
reduced doses for about as many *months* after the final disappearance of the disease, as it had existed *years* before. This
will prove the best security against a relapse. In plethoric
or inflamatory subjects the disease will yet be liable to relapse unless the diet be so regulated as to keep the system always free from increased vascular action. In some cases, stimulants must be entirely abandoned; in others, a sparing allowance of animal food appears to be essential to the preservation
of health; and in a few, vegetable diet for life. Cutaneous dis-
cases are sometimes complicated with diarrhœa, dyspepsia, or
general irritability of the stomach. Arsenic in small doses
will be found to soothe the bowels (the *pulse being quiet*) in
proportion as it allays the irritability of the skin. This assertion of our author, when first made, was treated with ridi-
cule; but after twenty years further observation he repeats it.
14

Arsenic, if rightly used, is adapted to the treatment of six out of every seven cases of chronic skin disease the physician is called upon to relieve. More than this, the diseases which are curable by arsenic are absolutely incurable without it, try what you will.

Our author gives the following specific directions for the use of Fowler's solution :

First. It should be given in divided doses—three doses in twenty-four hours, simply to avoid an unnecessarily large dose.

Second. It should be diluted with pure water, or if the case require the influence of antimony, the following should be ordered :

611.   ℞.   Liquoris potassæ arsenitis,   f.ʒij.
             Vini antimonii,   f.ʒxiv.
             Aquæ,   f.ʒj.   M.

A teaspoonful diluted three times a day.

Third. This dose should be taken with, or immediately *after*, a meal, in order that, being mixed with the patient's food, it may find a ready entrance into the blood, and that the bare possibility of its irritating the mucous membrane of the stomach or bowels may be avoided. Not that there is any danger of mischief, but the patient, aware that he is taking arsenic, may thus be disabused of all fanciful or imaginary suffering of this kind.

Fourth. It should be clearly understood that arsenic acts very slowly, and, therefore, it is best to begin with an average dose, say five minims of Fowler's solution, and this should be increased, not day by day, as was the practice thirty years ago, but two, three, or four weeks should be allowed to elapse before any necessity can exist for augmenting the dose. If, during this time, there should be set up an active or severe inflammation of the tunica conjunctiva in *both* eyes, the lower eyelid being swollen and showing on its lining membrane a horizontal streak of inflammation, then the medicine is not to be abandoned in a panic, but the dose may be reduced to four minims, and in a week or ten days the conjunctiva will be less inflamed ; or, if not, a lotion of the liquor plumbi subacetatis dilutus, or of cold black tea, will generally suffice to relieve the investments of the globe of the eye, and even to remove the slight degree of ecchymosis which is sometimes seen in the sclerotic tunic. But during this week, if the patient has been properly prepared for the course, the disease of the skin will show some amendment ; it will be shorn of its strength, and from this time the cure will be easy enough, although it may take many weeks, or even months, to effect it entirely. If it should be requisite that the course be continued for several months, the soles of the feet, and less frequently the palms of the hands, become more or less inflamed or rough, and very rarely slight vesications occur on the feet. These inconveniences must be borne patiently ; they will only exist during the course. Sometimes, also, after a protracted course, the skin of those parts of the body which are covered by the dress assume a dirt-brown, unwashed appearance, and under a lens present fine scales. This, also, is an ephemeral appearance, somewhat annoying to females of delicate complexion, but not for a moment to be compared with the afflictive form of disease which requires arsenic.

Fowler's solution should be used freshly prepared, for if kept very long, the bottle in which it is contained becomes lined with a very delicate film of metallic arsenic, impairing the strength of the solution.*

---

*Guide to the Treatment of Diseases of the Skin. Lond. 8th ed., p. 15; and Journal of Cutaneous Medicine for January, 1869, p. 350.

BALMANNO SQUIRE, M. B., F. L. S., ETC., LONDON.

612. ℞. Liquoris potassæ,   f.ʒss–iv.
      Aquæ.   f.ʒj.   M.

Useful locally in obstinate cases of lichen when the skin has become much thickened and is desquamating. The stronger solutions require great caution in their use.

Internally, small doses of *Donovan's solution* should be given. Cauterization with nitrate of silver is useful in some cases of persistent *lichen circumscriptus.**

ERASMUS WILSON, F. R. S., LONDON, ETC.

The constitutional treatment of lichen requires mild aperients, followed by bitters and mineral acids, by chalybeates and quinine. In chronic cases arsenic will generally effect a cure (F. 668.)

The local treatment of lichen calls for the use of ablutions with the juniper tar soap, tepid bathing, and anti-pruriginous and moderately stimulating lotions, such as F. 588, 589, 670.

But the most certain and powerful anti-pruriginous lotion is

613. ℞. Olei juniperi pyrolig.,
      Alcoholis,   aa. f.ʒj.
      Aquæ,   f ʒvj.   M.

This is very successful in *lichen urticatus.*†

STROPHULUS.

TILBURY FOX, M. D., LONDON, ETC.

In simple strophulus, cleanliness must be observed; the child must not be too much wrapped up; the use of soap must be avoided; the child should have proper food; the state

---

*A manual of the diseases of the skin. Lond., 1860, p. 34.
†Diseases of the Skin. Am. Ed., 1868, p. 188.

of health of the nerves should be seen to ; local irritation—
e. g., that of teething, hot clothing (flannel) must be remedied ; any aphthous state must be cured ; acidity should be
corrected, and a gentle aperient given ; tepid sponging,
spirit or alkaline lotions may be used locally. A very useful
one is

614. R. Soda carbonatis, Əj.
   Glycerinæ, f.ʒij.
   Aquæ rosæ, f.ʒvj. M.

Almond emulsion, lime water, and mild sulphur water may
also be used.

In the pruriginous form of strophulus, the patient must be
placed under the most favorable hygiene ; have good food,
good air, plenty of washing, and internally, iron, cod-liver
oil, and quinine, or chlorate of potash.*

PRURIGO.

TILBURY FOX, M. D., LONDON, ETC.

Speaking generally, there are three main objects in view in
the treatment of this affection ; the first, to improve the tone
of the general health ; the second, to allay the irritation of
the skin ; the third, to destroy any pediculi that may be present ; and this latter point must always be attended to.

615. R. Pulveris opii, gr.viij.
   Creasoti, ♏x.
   Adipis, ʒij. M.
   To allay irritation.

616. R. Sodæ carbonatis, ʒss.
   Succi conii,
   Aquæ sambuci, aa. f.ʒj.
   Useful in the earlier stages to allay itching. Or,

617. R. Sodæ carbonatis, ʒj.
   Glycerinæ, f.ʒjss.
   Aquæ sambuci, f.ʒvjss. M.

*Skin Diseases. Lond., 1869, p. 139.

Or,

618. ℞. Sodæ biboratis, ʒij.
Aquæ lauro-cerasi, f.ʒj.
Aquæ sambuci, f.ʒjss. M.

Either of the above are valuable anti-pruritic lotions.

619. ℞. Potassii cyanidi, gr.xv.
Aquæ, f.ʒviij. M.

This solution must be kept in a dark room. It will allay the cutaneous irritation.

620. ℞. Extracti belladonnæ, ʒss.
Acidi hydrocyanici dil., f.ʒss.
Glycerinæ, f.ʒj.
Aquæ, Oj. M.
Use diluted to relieve itching.

621. ℞. Tincturæ digitalis, f.ʒij–iv.
Glycerinæ, f.ʒss.
Aquæ rosæ, f.ʒvj. M.

This lotion is useful in prurigo of *purely neurotic character.* Or, the following:

622. ℞. Tincturæ nucis vomicæ, f.ʒij.
Tincturæ digitalis, f.ʒij–iij.
Glycerinæ, f.ʒij.
Aquæ rosæ, f.ʒvj–viij. M.

523. ℞. Plumbi acetatis, ʒj.
Acidi acetici diluti,
Aquæ destillatæ, aa. f.ʒij.
Olei olivæ, f.ʒiij. M.

624. ℞. Camphoræ, ʒss.
Alcoholis, f.ʒj.
Sodæ biboratis, Əij.
Aquæ rosæ, f.ʒviij. M.

The quantity of camphor may be increased in this anti-pruritic lotion, or camphor may be used in the form of

UNGUENTUM CAMPHORÆ.

625. ℞. Camphoræ, gr.x.
Glycerinæ, ♏x.
Adipis, ʒj. M.

626. ℞. Sodæ carbonatis, ℨij.
        Extracti opii, gr. x.
        Adipis, ℥ij. M.
    and add of
        Slacked lime, ℨj. M.

627. ℞. Sodæ hyposulphitis, ℨj.
        Glycerinæ, f.℥j.
        Aquæ, f.℥iij. M.

The above is particularly useful in *pruritus vaginæ.*

628. ℞. Caicii chloridi, ℥ss.
        Olei amygdalæ dulcis, f.℥ij.
        Adipis, ℥iij. M.
    To allay itching.

629. ℞. Liquoris potassæ arsenitis, ♏j.
        Vini ferri, f.℥j.
        Syrupi, f.℥ss.
        Aquæ destillatæ, f.℥ivss. M.
    A tablespoonful twice or thrice a day.

630. ℞. Sodæ arseniatis, gr. ij.
        Aquæ, q. s. M.

        to make a solution ; add

        Pulveris guiaici, ℨss.
        Hydrarg. oxysulphureti, Əj.
        Syrupi acaciæ, q. s. M.
    For xxiv pills. One, two or three times a day, in obsti-
    nate cases.

631. ℞. Extracti aconiti,
        Extracti taraxaci, aa. gr. xv. M.
    For xl pills. Two night and morning, in conjunction with
    starch baths and arsenite of iron in pills :

632. ℞. Ferri arseniatis, gr. iij.
        Extracti lupuli, ℨj.
        Pulveris altheæ, ℨss.
        Aquæ aurantii, q. s. M.
    For xlviij pills. One to two daily.

633. ℞. Extracti nucis vomicæ,   gr. ij.
       Fel. bovini purificati,   gr. vj.
       Extracti taraxaci,   gr. xxiv.
       Pulveris myrrhæ,   gr. xviij.   M.
For xxiv pills. One three times a day.*

634. ℞. Pulveris staphisagriæ,   ʒij.
       Olei olivæ,   f.ʒss.
       Adipis,   ʒss.   M.
Use in *prurigo pedicularis.*

PROF. HEBRA, OF VIENNA.†

Our author has employed the rubber-cloth treatment in cases of general itching of the skin (*prurigo senilis aut orum*) with the greatest benefit. The method of making and applying this cloth is described on page 223. He has a complete suit of rubber clothing made and worn next the skin, at first, day and night. Morning and evening the garments are either changed, or where, as in the hospital, only one suit is provided, this is removed for a short time, cleansed and again put on. The effect shows itself in each case the first day even ; perspiration becomes more abundant, the itching and tension of the skin entirely cease, and sleep, which had been disturbed, returns. All the patients expressed themselves as much relieved, and submit with great pleasure to the continued treatment. After the lapse of some weeks the rubber clothing is worn only either during the night or for an hour at a time during the day, until finally it is removed after complete cure in each case. The accompanying eczema is all the more easily treated by the ordinary remedies (tar preparations), as the coverings modify and mitigate the odor as well as conceal the stain they produce. To prevent misunderstanding, our author draws attention to the fact that the caoutchouc

---

*Skin Diseases. Lond., 1869, p. 369.
†Journal of Cutaneous Medicine, London, April, 1869, p. 41.

bandages are of themselves sufficient to allay the distressing itching, which compels the patients to scratch and deprive them of sleep, and the tar preparations are only used when sooner or later the ordinary symptoms of eczema show themselves.

BALMANNO SQUIRE, M. B , F. L. S., LONDON, ETC.

635. ℞. Hydrarg. chlor. corrosivi, gr. ij.
Aquæ, f.ℨj. M.

A lotion for *prurigo pubis.*

Or the surface may be dusted lightly with calomel, or the white precipitate ointment employed. The first effect of either of these applications is to increase the trouble, the moribund pediculi causing more irritation even than before, but in the course of two or three hours the irritation ceases altogether.*

ERASMUS WILSON, F. R. S., LONDON, ETC.

Arsenic, properly administered and watched, may be regarded as a specific in prurigo. Much may be accomplished toward the restoration of a healthy condition of the skin by ablutions with the juniper-tar and carbolic acid soap, frictions and manipulations with the hand after the manner of the shampooer, the tepid bath, the sweating bath used with discretion, and moderately stimulating local applications.†

A local remedy frequently of service in allaying the itching of *prurigo senilis* is glycerine applied with a sponge.

ANTI-PRURITICS.

The best applications suited for the temporary relief of pruritus are vinegar, lemon juice, weak solution of corrosive sublimate, tincture and watery solution of opium, creasote ointment and lotion, tar ointment, and especially that of juniper

*A Manual of Diseases of the Skin. Lond., 1859, p. 42.
†Diseases of the Skin. Am. Ed., p. 302.

tar (F. 537,) ointment of opium with camphor, the diluted nitrate of mercury ointment, ointment of lime (F. 626), ointment of cyanide of potassium (F. 647), lotion of hydrocyanic acid (F. 670), aconite, acetate of ammonia, muriate of ammonia, sulphuret of potash, chlorate of soda, etc.

The following formulæ are all useful :

636.	R.	Calcis hydratis,	ʒij.	
		Sodæ carbonatis,	ʒss.	
		Tincturæ opii,	f.ʒss.	
		Adipis,	ʒj.	M.
637.	R.	Tincturæ opii,	f.ʒss.	
		Sulphuris sublimati,	ʒss.	
		Zinci oxidi,	ʒj	
		Olei amygdalæ dulcis,	f.ʒj.	
		Adipis,	ʒiij.	M.
638.	R.	Hydrarg. sulphureti rubri,	ʒij.	
		Tincturæ opii,	f.ʒij.	
		Sulphuris sublimati,	ʒss.	
		Adipis,	ʒv.	M.
639.	R.	Ammoniæ muriatis,	ʒj.	
		Pulveris hellebori albi,	ʒss.	
		Adipis,	ʒiij.	M.

For local prurigo.

It is well to have at command a number of anti-pruritic remedies, for those that succeed in one case will often fail in others. The above list furnishes a number to select from.

VESICULAR DISEASES.

ECZEMA.

EDGAR A. BROWNE, SURGEON TO LIVERPOOL DISPENSARY FOR SKIN DISEASES.

640.	R.	Acidi carbolici fluidi,	f.ʒss-j.	
		Unguenti zinci oxidi,	ʒj.	M.

This ointment is useful in hardening the newly-formed epidermis. In the latter stages of the disease it may be used instead of the tarry preparations. The facility with which its strength may be graduated to suit the varying susceptibility of the skin in various cases renders it a convenient application in a disease which varies so much in severity as eczema.*

J. M. DA COSTA, M. D., PHILADELPHIA.

641. R. Potassii cyanidi,   gr.¼.
          Alcoholis,   f.ʒij.
          Glycerinæ,   f.ʒss.
          Aquæ,   f ʒvj.   M.

A local application to allay itching in various skin affections, to be sponged over the part several times a day.

642. R. Hydrarg. chlor. mitis,   ℈j.
          Cerati simplicis,   ʒj.   M.

An alterative ointment to be applied in *eczema capitis*, after poulticing.

In the treatment of *eczema impetiginoides* apply poultices to get rid of the crusts and then use

643. R. Potassæ carbonatis,   ʒj.
          Aquæ,   Oj.   M.

The parts are to be kept constantly enveloped in this lotion, day and night.   Internally,

644. R. Liquoris potassæ arsenitis, f.ʒss–jss.
          Vini ferri amari,
          Aquæ,   aa. f.ʒjss.   M.

A tablespoonful ter die, after meals.

After the eruption subsides, use the following stimulating ointment, occasionally applying a poultice if necessary to remove the crusts:

645. R. Ung. hydrargyri nitratis,
          Ung. picis liquidæ,
          Cerati simplicis,   aa. ʒij.
          Glycerinæ,   f.ʒij.   M.

Strong tar water is also useful.

____
*The Practioner. London, Dec. 1869, p. 335.

TILBURY FOX, M. D., LONDON, PHYSICIAN TO THE SKIN DE-
PARTMENT, CHARING-CROSS HOSPITAL, ETC., ETC.

646. R. Zinci oxidi,
         Calaminæ preparatæ,   aa. ℥j.
         Glycerinæ,   f.℥iss.
         Aquæ rosæ,   q. s. ad. f.℥vj.   M.

Use in eczema, generally when the surface is tender and
red. The part should be lightly bandaged with this lotion,
which should be used very freely so as to keep the surface
moist and exclude the air if possible. If the itching or sensation of burning is bad, the following may be used:

647. R. Potassii cyanidi,   gr.iij-v.
         Adipis,   ℥j   M.

In the second, or exudative stage, ointments should be generally avoided. In proportion as the heat or itching, the redness or swelling disappear, astringents should be employed;
but whenever there are signs of irritation, soothing and emollient remedies should be used externally. This treatment,
together with aperient tonics, generally controls the discharge.
The diseased parts should be most gently handled at all
times. Soap should not be used, and no friction with the
clothes allowed. When the third or scaly stage is reached, it
is often still highly necessary to avoid the use of any application which acts as an irritant, for irritability is one of the
chief characteristics of the skin of an eczematous subject.

Astringents are generally called for in simple forms of
eczema, such as is seen in the scalp. Our author prefers, in
connection with tonics, the use at the outset of

648. R. Sodæ biboratis,   ℈ij.
         Plumbi acetatis,   gr.ij.
         Glycerinæ,   f.℥j.
         Adipis,   ℥j.   M.

A stronger ointment is

649. ℞. Ung. hydrag nitratis, ℨij.
    Glycerinæ, f.℥ij.
    Adipis, ℥ij.   M.

Where thickening and induration finally remain, these may be regarded as secondary and ordinary results of congestion, and should be treated accordingly by revulsives. Our author often uses

650. ℞. Argenti nitratis, ℈ij.
    Ætheris nitrici, f.℥j.   M.

Or,

651. ℞. Olei juniperis pyrolignei, f.ℨj-iij.
    Adipis, ℥j.   M.

Should this not suffice, order

652. ℞. Hydrargyri iodidi rubri, gr.v-xv.
    Adipis, ℥j.   M.

The above line of procedure holds good in the case of children ; but here in addition an absorbent powder is serviceable. It may be

653. ℞. Zinci oxidi,
    Calaminæ preparatæ,
    Amyli, aa ℥ss.   M.

Our author prefers a lead or calamine lotion, with exclusion of air, and at night a layer of elder-flower ointment, to anything else as simple applications in *eczema infantilis.*

654. ℞. Pulveris aluminis, ℨij.
    Infusi rosæ, Oj.   M.
Used in *eczema sine crustis.*

655. ℞. Potassæ cyanidi, gr.v.
    Sulphuris,
    Potassæ bicarbonatis, aa. ℨss.
    Cocci cacti, gr.j.
    Adipis, ℥j.   M.
In eczema with pruritis.

656. ℞. Camphoræ,   ʒss.
      Alcoholis, q. s. to dissolve ;
      add
      Zinci oxidi,
      Amyli,   aa. ʒss.   M.

Use as a powder to allay the *burning heat of eczema.*

657. ℞. Camphoræ,   gr.viij.
      Tincturæ conii,   f.ʒij.
      Cerati adipis,   ʒj.   M.

658. ℞. Saponis mollis,   ʒj.
      Aquæ bullientis,   Oj.   M.

Scent with some essential oil and use in the second stage of
eczema to counteract the infiltration.

659. ℞. Saponis mollis,
      Alcoholis,
      Olei cadinii,   aa. f.ʒj.
      Olei lavandulæ,   f.ʒiss.   M.

This preparation is more elegant than Hebra's "Tr. saponis viridis cum pice." (F.665.)

660. ℞. Olei juniperis pyrolignei,   f.ʒj–viij.
      Adipis,   ʒj.
    Mix with ʒss of mutton suet.

661. ℞. Picis liquidæ,   f.ʒj.
      Camphoræ,   gr.x.
      Adipis,   ʒx.   M.

662. ℞. Liq. potassæ arsenitis,   ♏lxxx.
      Potassii iodidi,   gr.xvj.
      Iodinii,   gr.iv.
      Aquæ aurantii floris,   f.ʒij   M.

A tablespoonful ter die.

PROF. S. D. GROSS.

663. ℞. Tr. ferri chloridi,   f.ʒj.
      Liq. potassæ arsenitis,   f.ʒiss.
      Hydrarg. chlor. corros.,   gr iij.   M.

Thirty drops ter die, in sweetened water, through a tube, as
an alterant tonic for eczema.

664. R. Saponis mollis, ℥j.
Aquæ bullientis, Oj. M.

Use in the second stage of *eczema*, to counteract the infiltration. The lotion may be scented with some essential
oil.

665. R. Picis liquidæ,
Alcoholis,
Saponis mollis, aa. f.℥ij. M.

This constitutes the "tr. saponis viridis cum pice," and is
used in *eczema*. (See F. 659.)

Our author has of late frequently employed *rubber-cloth* in
the treatment of every variety of this affection from *eczema
squamosum* to *eczema impetiginosum*. He makes use either of
closely applied pieces, roller bandages, or of whole garments made of this material. The rubber cloth (*toile caout-
chouque*) consists of ordinary cotton, which is first coated with
a solution of caoutchouc, and then submitted to the process
known as vulcanizing. This consists in sprinkling the stuff
with a mixture of caoutchouc and sulphur, and exposing it to
a high temperature under a pressure of sixteen atmospheres.
The material obtained in this way is gray, black, or of any
other desirable color, flexible, impermeable to watery fluids,
smooth and polished on one surface, dull and rough on the
other, and smells of caoutchouc and sulphur. Oil, as well as
fats, and alcohol dissolve this layer of caoutchouc, thus destroying its desirable qualities, and rendering it useless for the
purpose in question. It can be worked like any other cloth,
that is, be cut, sewed, and its surfaces be made to adhere by
means of a cement containing caoutchouc. These properties
led our author not only to apply it simply to the affected
parts, but to have various pieces of clothing made of it, for
instance, caps for the head, bags in which to envelop various

regions of the body, gloves, stockings, and finally, entire
drawers, with and without feet attached, as well as shirts and
blouses.   Besides these, he has some of the ordinary gum-
elastic (not vulcanized) made into bandages and gloves, and
is convinced that this is also useful.

In every case the smooth side of the vulcanized cloth is laid
in contact with the skin, from which the collections of morbid
products, the scales, crusts, etc., have been previously removed,
although in some cases the cloth is applied, for the sake of
experiment, above these.   On removing the cloths, at the end
of 12 or 14 hours, they are found very moist, often entirely
soaked through, and the fluid, which had collected on the surface of the skin in considerable quantity, of a penetrating
smell, worse even than that of the " stinking foot-sweat."   The
skin itself, however, when cleansed from the diseased products
thus softened, appears odorless and only reddened, more or less
robbed of its epidermis, moist and shiny.   The sensations of
the patient during their application are not at all unpleasant,
there being no pain or itching.   After their removal itching
generally comes on, and, unless they are renewed within half
an hour or so, a feeling of contraction and pain also, so that
the patients long for their immediate reapplication.   If the
treatment is continued in this way, the whole series of symptoms gradually diminish—the moistening, redness, itching, and
pain—and in many cases the cure of the eczema is seen to be
complete in the course of two months.

But as it is known that under other treatment the cure of
eczema may be effected in this period of time, the question
arises what advantage the caoutchouc method offers over
others, such as by ung. diachyli, tar, zinc, sublimate, etc.

The answer must be that, although in general no excessive
advantage can be attributed to the caoutchouc, nevertheless.

there are cases in which this new remedy can be used with especial profit. It applies particularly to eczema of the hands, fingers, flexures of the joints, scrotum and feet, in which the application of salves, etc., is not only attended by much inconvenience to the patient, but in which also the caoutchouc preparations are able to afford a much more speedy relief to the pain produced by the fissures, inasmuch as such parts can be kept constantly moist by the easy application of the gloves, coats, bandages, suspensories and stockings employed. Although, therefore, no new panacea has been introduced into dermato-therapeutics by the use of caoutchouc in the treatment of eczema, it must still be regarded as a very *valuable addition to our means of cure*, and all the more as it does not prevent the helping use at the same time of other known remedies. Thus in many cases the cure of eczema is powerfully assisted by the simultaneous use of schmierseife, baths, douches, tar preparations, etc., and these latter in turn made more serviceable by the application of the caoutchouc clothes.*

FRANK F. MAURY, M. D., PHILADELPHIA.

666. ℞. Ung. hydrarg. nitratis, ℧iss.
 Olei olivæ, f.℧iss.
 Glycerinæ, f.℧iss. ℞.
 A pomade in eczema, etc.

PROF. JOSEPH PANCOAST, PHILADELPHIA.

667. ℞. Zinci oxidi, ℧iss.
 Amyli, ℧j.
 Cerati adipis, ℧ss.
 Glycerinæ, f.℧ss. M.
 For application to ulcers, eczema, etc.

---

*Journal of Cutaneous Medicine, London, April, 1869, p. 41.

ERASMUS WILSON, F. R. S., LONDON, ETC.

FERRO ARSENICAL MIXTURE.

668. R. Liq. potassæ arsenitis, f.ʒss–j.
Vini ferri, f.ʒjss.
Syrupi, f.ʒiij.
Aquæ anethi, f.ʒij. M.
A teaspoonful ter die for *eczema infantilis.*

No better formula can be employed for the administration of Fowler's solution to infants. It should be given on a full stomach. Our author considers it a specific in this affection. For infants of two years and under, the weaker solution (representing one minim of Fowler's at a dose) should be employed. For two to seven years the stronger solution (equal to two minims of Fowler's at a dose).

Locally, apply the benzoated ointment of oxide of zinc, rubbed down with alcohol, in the proportion of a drachm of the latter to an ounce of the former. The eruption is to be completely covered with this ointment, which should be applied night and morning, and oftener if accidentally displaced during the day. It should be kept on as a permanent dressing, a piece of thin flannel or linen rag, a sheet of cotton-wool, or a slip of tissue paper being laid over it.

When the eruption covers more or less of the entire body, a little shirt made of old linen, with sleeves for the arms and legs, with means for fastening around the legs, and if necessary over the hands and feet, is to be provided and worn constantly day and night, for a week together, if requisite. When the eruption is confined to the arms or legs, linen sleeves or an elastic cotton bandage will be sufficient. On the face, no other covering than the ointment is necessary; sometimes, in this situation, small pieces of tissue paper may be laid over the ointment with advantage. In this way the formation of crusts is prevented, or they are dislodged if already formed.

When the eruption passes from the acute to the chronic state, gentle friction of the skin, with the ointment, is desirable and grateful to the infant.

On the scalp, the ointment should be applied in the direction of the hair to avoid matting. So soon as the ichorous discharge has lessened, the hair should be gently brushed. When the ointment accumulates too thickly over a given part so as to confine the secretions, the whole of it should be carefully washed off the part with the yolk of an egg, and, after drying the skin, fresh ointment substituted. Otherwise the ointment should not be disturbed.

The inflamed skin should never be washed; it may be wiped with a soft napkin to remove exudations or secretions.

The use of benzoated ointment of oxide of zinc, and of the ferro-arsenical mixture, renders a failure impossible in the treatment of eczema infantilis. This medication should always be preceded by the administration of

669. ℞. Hydrarg. chlor. mitis,
Sacchari albi, aa. gr.j. M.

For one powder, to be repeated according to circumstances, once a week (which is usually sufficient), twice a week, every other night for a few times, or even every night for two or three nights, if absolutely necessary.

In the transitional and passive periods of eczema, pruritus is extremely troublesome and demands care. As an antipruritic, the following is an admirable cooling and soothing mixture:

670. ℞. Acidi hydrocyanici dil., f.ʒij.
Alcoholis, f.ʒxiv.
Emul. amygdalæ amaræ, f ℥vj. M.

(The emulsion of bitter almonds contains twenty or thirty kernels to f.℥vj. of water.)

All lotions, however agreeable at first, are apt, unless they
contain oil or glycerine, to leave behind them a certain degree
of dryness, or, perhaps, add to the dryness they were intended
to mitigate, for dryness of itself may be an incidental cause
of pruritus. Hence, as soon as the lotion is dried, a smear with
the benzoated oxide of zinc ointment should follow.*

HERPES.

TILBURY FOX, M. D., LONDON, PHYSICIAN TO THE SKIN
DEPARTMENT, CHARING-CROSS HOSPITAL, ETC.

671. R. Acidi carbolici,   ʒij.
          Glycerine,   f.ʒj.
          Aquæ rosæ,   ad f.ʒviij   M.
      Use in ring-worm of the surface especially.

SKIN HOSPITAL, LONDON.

672. R. Zinci carbonat præcipitati,   ʒj.
          Liq plumbi subacetatis,   ♏x.
          Acidi hydrocyanici diluti,   ♏xx.
          Glycerinæ,   ♏xx.
          Adipis,   ʒj.   M·
                                                     †

PUSTULAR DISEASES.

ACNE.

J. M. DA COSTA, M. D., PHILADELPHIA.

673. R. Acidi carbolici fluidi,   ♏·xxx.
          Glycerinæ,   f.ʒij.
          Cerati adipis,   ʒvj.   M·

Employed in the treatment of acne and other pustular skin
affections, in some cases with signal effect. If it produce too

---

*Journal of Cutaneous Medicine, London, Oct. 1869, p. 225.
†Squire's Pharmacopœias of the London Hospitals. London, 1869, p. 174.

much irritation in this strength, it may be diluted with fresh
lard.

675. R. Liquoris potassæ arsenitis, f.ʒj.
Extracti cascarillæ fluidi,
Tincturæ rhei dulcis, aa. f.ʒx. M.

A teaspoonful ter die. Locally, iodide of sulphur ointment
(gr.xv to adeps ʒj) twice a day, in chronic cases.

In simpler cases, try first a very mild ointment. None is
more soothing than one of lead :

675. R. Liquor. plumbi subacetatis, ℔xx.
Glycerinæ, f.ʒj.
Cerati simplicis, ʒvij. M.

To be rubbed on thoroughly, morning and evening.

After arsenic, cod-liver oil is a good remedy. Iron may
succeed in cases in which cod-liver oil and arsenic have failed.

In treating acne in women, it should be borne in mind that
it, more than any other affection of the skin, bears a relation
to uterine disorders.

TILBURY FOX, M. D., ETC., LONDON.

In the treatment of acne, it is necessary, first of all, to insure
cleanliness; secondly, to remove any cause of debility present,
correct menstrual deviations, cure dyspepsia, etc., and especially to prevent constipation. These preliminary cares are
sine qua non to success. Then, in the simpler cases, which
exhibit little inflammatory action, friction and gentle stimulation may be had recourse to; borax, soda, and calamine
lotions, or the following will suffice :

676. R. Hydrarg. chlor. corrosivi, gr ij.
Emul. amygdalæ amaræ, f.ʒviij. M.

In the severe forms much more remains to be done. The
general condition of the health must be improved, and whatever special indications which are present be fulfilled. Locally, if there be much inflammation, warm poultices, hot

vapor douches, poultices, and warm lead lotions are called for.
When these have allayed the irritation, absorbents may be
used—oxide of zinc lotion or the oxide of zinc and glycerine.
Our author generally prescribes

677. R. Hydrarg. chlor. corrosivi, gr.ij.
          Sodæ biboratis,   ℈ss.
          Glycerinæ,   f.ʒj.
          Aquæ,   f ℥vij.   M.
     To be frequently used.

When the disease is chronic revulsives are needed. The
following is one of the best:

678. R. Hydrargyri iodidi rubri,   gr. v.
          Aquæ,   f.ʒj.   M.
     A very good plan is to pencil each spot with acid nitrate of
     mercury, once or twice.

In *acne rosacea*, diet and good hygiene are of vast importance. If there be many varicose vessels, they may be cut
across—the incisions never being deeper than two lines. Cold
water will stay the bleeding and collodion may be subsequently used to.contract and heal the incisions. Acids and
pepsin given internally do much good. Much has been said
of the efficacy of the iodo-chloride of mercury in *acne roacea*
and *indurata*. The following formula is used:

679. R. Hydrargyri iodo-chloridi, gr.v–xv.
          Adipis,   ʒj.   M.

The ointment requires care, as it produces a good deal of
irritation.

LOTIO BORACIS COMPOSITA.

680. R. Sodæ biboratis,
          Ammoniæ carbonatis, aa. ʒjss.
          Acidi hydrocyanici diluti, f.ʒiij.
          Glycerinæ,   f.ʒj.
          Aquæ destillatæ,   Oj.   M.
     To be mixed when used with one, two or four times its
     bulk of water.

681. ℞. Sodæ biboratis,   Ӡj.
          Zinci oxidi,   ʒj.
          Liq plumbi subacetatis,   f.ʒij.
          Aquæ calcis,   f.ʒvj-viij   M.

682. ℞. Hydrarg. iodidi viridi,   gr.ij-xv.
          Adipis,   ʒj.   M.

683. ℞. Hydrargyri bicyanidi,   gr.v-x.
          Adipis,   ʒj.   M.

In *acne indurata.* Or,

684. ℞. Hydrarg. iodidi rubri,   gr.v-xx.
          Adipis,   ʒj.   M.

685. ℞. Sulphuris iodidi,   gr.x–ʒj.
          Adipis,   ʒj.   M.

686. ℞. Phosphori,   gr.ij–v.
          Ætheris,   q. s.
          dissolve and add
              Camphoræ,   Ӡj.
              Cerati simplicis,   ʒss.   M.

687. ℞. Hydrarg. chlor. corros.,   gr.j.
          Tincturæ benzoini,   f.ʒij.
          Aquæ destillatæ,   f.ʒvj.   M.

688. ℞. Iodinii,   gr.j.
          Olei amygdale dulcis,   f.ʒj.
          Olei oliviæ,   f.ʒij.   M.

A tablespoonful ter die in the acne of scrofulous persons.

689. ℞. Phosphori,   gr. v.
          Ætheris,   f.ʒj.   M.

Five to ten minims ter die.

690. ℞. Hydrargyri iodo-chloridi,   gr.iv.
          Panis,   ʒijss.
          Syrupi,   q. s.,   M.

For 100 pills. One to three daily.

691. ℞. Phosphori,   gr.iij–xx.
          Olei amygdalæ dulcis,   ♏x-lx.
          Pulveris acaciæ,   q. s.   M.

For cxii pills. One twice a day.*

* Skin Diseases. London, 1869, p, 400.

PROF. HEBRA, OF VIENNA.

Our author treats acne as follows : He gives vapor douches
to the face, applies soft soap or

692. R. Potassæ causticæ,   ℨj.
             Aquæ,   Oj.   M.

In other cases he washes the face with soft soap, and at night
applies a paste made as follows :

693. R. Sulphuris,   ℨj.
             Alcoholis,   f.ℨj.   M.
To be painted on by means of a camel hair pencil. This is
    removed in the morning by means of soap. Cocoa butter
    is kept on all day.

He sometimes uses

694. R. Hydrarg. chlor. corrosivi, gr.v.
             Alcoholis,   f.ℨj.   M.
To be applied with a compress for two hours.

At other times he applies two or three times a day

695. R. Hydrarg. chlor. corrosivi, gr.j.
             Tincturæ benzoini,   f.ℨj.
             Aquæ,   f.ℨvj.   M.

ALEXANDER THOMPSON, M. D., MOUNT SAVAGE, MD.

696. R. Hydrargyri chlori. corrosi., gr.ij.
             Potassii iodidi,   Эj.
             Sodæ biboratis,   ℨss.
             Spirit. ammonia aromatici, f.ℨss
             Aquæ cologni,   f.ℨss.
             Aquæ camphoræ,   f.ℨiij.   M.
For acne indurata, to be applied with a fine sponge once or
    twice daily.

Our author, who has had a large experience with this affection, speaks highly of the effects of this local treatment in
chronic cases.

IMPETIGO.

J. M. DA COSTA, M. D.

697. ℞. Unguenti picis,
Ung. hydrar. ox. rubri, aa. ℥ss.

For *impetigo.* To be rubbed in morning and night.

If this fails, apply

698. ℞. Cupri sulphatis,   ℈j.-ij.
Aquæ,   f.℥j.   M.

Or use the solid sulphate of copper.

TILBURY FOX, M. D., LONDON, M. R. C. P., ETC.

699. ℞. Plumbi acetatis,   gr.xv.
Acidi hydrocyanici dil.,   ℳxx.
Alcoholis,   f.℥ss.
Aquæ,   aa. f.℥vj.   M.

Use in *impetigo.*

At the outset of this disease, and in direct proportion to the degree of irritation present, remedies should be emolient in character. Poulticing is the first step. Then the above lotion may be used. Subsequently,

700. ℞. Hydrargyri ammoniati,   ℈j.
Olei olivæ,   f.℥j.
Adipis,   ℥j.
Olei rosæ,   ℳvj.
Tincture tolutanus,   gtt.xx.   M.

If the scalp is affected, the hair must be cut from around the disease. Pediculi, if they exist, will be destroyed by the above white precipitate ointment. In many cases alkaline lotions are of use, for example,

701. ℞. Sodæ carbonatis,   ℥j.
Aquæ,   f.℥vj.   M.

ECTHYMA.

DR. FELIX VON NIEMEYER, PROFESSOR UNIVERSITY OF TUBINGEN.

The only species of ecthyma which requires active treatment is the chronic variety, with tendency to ulceration of the skin. First of all, the cachectic condition must be corrected, if possible, by means of proper ventilation, generous food, the use of wine, good beer, and the preparations of iron and quinine. Externally, while the inflammation continues, warm poultices must be applied; afterward, when the ulcers are indolent, they require stimulation, especially by touching their surface with nitrate of silver.*

RUPIA.

DR. FELIX VON NIEMEYER, PROF. UNIVERSITY OF TUBINGEN.

The principal task in the treatment of rupia consists in combating the constitutional vice upon which it depends. If successful, young epidermis soon forms beneath the scabs, and the ulcers heal. But if the attempt to improve the constitution fails, local treatment will generally be found useless also. Besides the constitutional remedies, the scabs may be softened by poulticing. The ulcers which remain require stimulating applications, such as repeated touching with lunar caustic.†

SKIN HOSPITAL, LONDON.

702. R. Syrupi ferri iodidi,   f.ʒjss.
          Magnesiæ sulphatis,   ʒj.
          Olei menthæ piperitæ,   gtt.ij.
          Aquæ,   ad f.ʒiv.   M.
     Dose—f.ʒij-f.ʒiv; in pustular and sebaceous affections.‡

---

*Text-Book of Practical Medicine, vol. ii., p. 436.
†Text Book of Practical Medicine. Am. ed., vol. ii., p. 439.
‡Squire's Pharm. of the London Hospitals. London, 1869, p. 96.

SQUAMOUS DISEASES.

LEPRA.

J. M. DA COSTA, M. D., PHILADELPHIA.

703. ℞. Sodæ sulphitis, ℥ss.
Aquæ, f.℥vj. M.

To be used as a wash in *lepra*.

The patient at the same time being ordered, internally

704. ℞. Liq. potassæ arsen., ℳiij.
Tr. gentianæ comp., f.℥ij. M.

For one dose ter die.

PSORIASIS.

DR. M'CALL ANDERSON, OF GLASGOW.

705. ℞. Acidi carbolici cryst. ʒjss.
Glycerinæ, q.s.
Aquæ destillatæ, f.℥vj. M.

A teaspoonful in a wine-glass of water, three times a day on an empty stomach.

J. M. DA COSTA, M. D., PHILADELPHIA.

706. ℞. Ung. hydrarg. oxidi rubri,
Ung. hydrargyri, aa. ʒij.
Glycerinæ, f.℥ss. M.

For psoriasis; to be rubbed in morning and evening, when there are no vesicles, after washing the parts with castile soap.

Internally,

707. ℞. Liq. arsenici et hydrargyri
iodidi, f.℥ss.
Ext. dulcamaræ fluidi, f.℥ijss. M.

A teaspoonful ter die, after meals.

Avoid fatty articles of diet and those highly salted. The most important thing in skin diseases is to determine, not so much their character externally, as to ascertain with what internal conditions they are associated.

In the acute stages of psoriasis, the following may be used:

708. R. Cerati plumbi subacet.,   ʒij.
          Glycerinæ,   f.ʒj.
          Cerati simplicis,   ʒiv.   M.

Attention should be paid to the digestive system. Then, after the acute inflammatory condition has subsided, the red precipitate ointment (F. 706) may be employed, or

709. R. Sulphuris iodidi,   gr.x.
          Adipis,   ʒj.   M.
To be rubbed in morning and evening.

Or,

710. R. Ung. hydrargyri nitratis,
          Unguenti picis,
          Cerati adipis,   aa. ʒss.   M.

Internally, Donovan's solution combined as above. (F.707.)
Or,

711. R. Liq. arsenici et hydrarg.
          iodidi,   f.ʒij.
          Tinct. cinchonæ comp,   f.ʒiij.   M.
A desertspoonful ter die.

Carbolic acid soap is also a very useful article. It is one of the nicest ways of using carbolic acid. It is a new preparation, hardly got into general practice here. Our author uses it quite largely in skin affections, and thinks very favorably of it.

Dr. ERASMUS WILSON speaks highly of the carbolic acid soap recently manufactured by Mr. CRACE CALVERT. It contains twenty per cent. of carbolic acid and an equal quantity of glycerine. It is transparent and elegant in appearance. He has used it for several months with increasing satisfaction. It gives a freshness to the skin unequaled by any other soap.*

---

*Journal of Cutaneous Medicine. London, Oct., 1869, p. 284.

TILBURY FOX, M. D., LONDON, PHYSICIAN TO THE SKIN
DEPARTMENT, CHARING-CROSS HOSPITAL.

712. R. Argenti chloridi,   gr.v–xv.
          Cerati adipis,   ʒvj.   M.
A useful ointment in this affection.

If the disease is slight and localized to a few spots only,
treatment may be commenced at once with tarry applications,
for the scales are thereby removed sufficiently well.

713. R. Olei juniperis pyrolignei,   f.ʒij.
          Olei olivæ,   f.ʒj.
          Adipis,   ʒj.   M.
To be used night and morning. Or,

714. R. Creasoti,   gtt.vj.
          Unguer.ti hydrargyri,   gr.xv.
          Adipis,   ʒij.   M.

Where the disease is more extensive, or the scales thickly
covering the patch, alkaline baths are to be employed (four
ounces of carbonate of soda to each bath,) the patient soaking
for some twenty minutes or so ; or individual patches may be
softened up with water-dressing, or glycerine plasma, and this
is especially necessary in hardened spots about the hands and
feet. Independently of its softening action, the alkaline
bathing seems to exert some curative power. When the
scales have ceased to form freely, use

715. R. Argenti nitratis,   ℈ij.
          Ætheris,   f.ʒj.   M.
This painted over the spots, night and morning, helps the
    cure.

In chronic cases, with thickening of the patches, or where
there is much elevation of the disease, as in the *nummular*
variety, a more decided impression may be produced by

716. R. Picis liquidæ,
          Alcoholis,   aa. f.ʒij.   M.
To be rubbed in with flannel.

When there is a tendency to "discharge," use

717. R. Ung. hydrargyri nitratis, ℨij.
            Glycerine,   f.ℨij.
            Linimenti camphoræ,   f.ℨj.   M.

When cracking occurs, as in the palmar and plantar varieties, a paste made of glycerine and borax is useful; or the cracks may be touched with nitric acid.

The ill success which attends the treatment of this affection is generally due not to a want of remedies, but to the mode of their application and an unattention to individual peculiarities of diathesis and derangement in the assimilative and secreting organs.*

SKIN HOSPITAL, LONDON.

718. R. Ammoniæ muriatis,   Əj.
            Unguenti hydrargyri,   ℨj.
            Olei amygdalæ amaræ, ℩iv.
            Adipis,   q. s. ad ℨj.   M.
Use in squamous and tubercular affections.†

719. R. Creasoti,   ℩vj.
            Hydrargyri oxidi rubri, gr.x.
            Plumbi carbonatis,   Əj.
            Adipis,   ℨj.   M.
Half an ounce of *palm oil* may be added to this, and also the following: ‡

720. R. Plumbi acetatis,   gr.x.
            Zinci oxidi,   Əj.
            Hydrarg. chlor. mitis,   gr.x.
            Ung. hydrarg. nitratis, Əj.
            Adipis benzoati,   ad ℨj.   M.
Used in squamous and ulcerous affections.§

721. R. Ung. hydrarg. nitrat's, ℨij.
            Adipis,   ℨvj.   M.
Used in squamous and parasitic affections.‖

---

*Skin Diseases, London, 1869, p. 200.
†Squire's Pharm. of the London Hospitals. London, 1869, p. 183.
‡Ibid., p. 175.
§Ibid., p. 177.
‖Ibid., p. 178.

SKIN HOSPITAL, LONDON.

722. ℞. Bismuthi subnitratis,   gr.vj.
    Hydrarg. chlor. cor.,   gr.ss.
    Spiritûs camphoræ,   ℩ iss.
    Aquæ,   ad f.℥j.   M.

Use diluted with 1 to 3 parts of water, in squamous, pustular, vesicular, and sebaceous affections.

723. ℞. Ammoniæ phosphatis,   ℥iss.
    Ammoniæ carbonatis,   ℈ijss.
    Spir. lavandulæ comp.,   ℩xij.
    Aquæ,   ad f.℥iv.   M.

Used in doses of f.℥ij to f.℥iv, in squamous and papular affections.

724. ℞. Cupri ammoniati,   gr.ij
    Spiritûs am. aromatici,
    Tincturæ hyoscyami,   aa. f.℥iss.
    Aquæ,   ad f.℥iv.   M.

Used in doses of f.℥ij—f.℥iv. in squamous affections.

725. ℞. Tincturæ ferri chloridi,   f.℥ijss.
    Acidi arseniosi,   gr.iij–v.
    Acidi muriatici,   ℩.v.
    Aquæ,   ad f.℥iij.   M.

Used in doses of f.℥ij-f.℥iv, in chronic, squamous, and pustular affections.

726. ℞. Hydrarg. chloridi cor.,   gr.iss.
    Acidi muriatici,   ℩.iss.
    Acidi arseniosi,   gr.ij–v.
    Ammoniæ muriatis,   gr.iij.
    Spiritûs lavandulæ comp.,   ℩ vj.
    Aquæ,   ad f.℥ijss.   M.

Used in doses of f.℥j to f.℥ij, in squamous and cachectic affections.

727. ℞. Magnesiæ sulphatis,   ℥vj.
    Magnesiæ carbonatis,   gr.xxxvj.
    Tincturæ colchici,   f.℥iss.
    Aquæ menthæ piperitæ, ad f.℥iij.   M.

In acute squamous and vesicular affections, in doses of f.℥ij-f.℥iv.

728. ℞. Acidi arseniosi,   gr.1-20.
          Extracti jalapæ,   gr.11-5.
          Pulveris aromatici comp.,   gr.14-5.
          Pulveris acaciæ,   gr.3-10.
          Glycerinæ,   q. s.   M.

For one pill, twice a day in squamous affections.

729. ℞. Hydrarg. chloridi cor.,   gr.1-10.
          Extracti aconiti,   gr.
          Extracti conii,   gr.   M.

For one pill, twice a day in squamous and cachectic affections*.

PITYRIASIS.

J. M. DA COSTA, M. D., PHILADELPHIA.

730. ℞. Ung hydrarg. nitratis,
          Cerati simplicis,   aa. ℥ss.   M.

For *pityriasis of the scalp.* To be applied morning and night.
The hair should be cut short, and poultices applied before
using this ointment. The scalp is to be kept clean with
soap.

TILBURY FOX, M. D., LONDON, PHYSICIAN TO THE SKIN DEPARTMENT, CHARING-CROSS HOSPITAL.

731. ℞. Creasoti,   gtt.xl.
          Glycerinæ,   f.℥iij.
          Aquæ,   f.℥vj-viij.

Used in pityriasis.

732. ℞. Hydrargyri ammoniati,   Əj.
          Olei olivæ,   f.℥j.
          Adipis,   ℥j.
          Olei rosæ,   ♏vj.
          Tincturæ tolutanus,   gtt.xx.   M.

Use in *pityriasis capitis.*

THOMAS HAWKES TANNER, M. D., F. L. S., LOND.

733. ℞. Ung creasoti,
          Ung. sulphuris,   aa. ℥ss.   M.

In pityriasis and some other chronic cutaneous affections.

---

*Squire's Pharm. of the London Hospitals. London, 1869, pp. 79.

ICHTHYOSIS.

TILBURY FOX, M. D., ETC., LONDON.

Our author is pretty successful in getting patients into a
continuously comfortable condition. In the first place, he is
careful to see that patients are cleanly ; that they are well fed
and clothed. He then gives cod-liver oil and such remedies
as quinine. He does not prescribe arsenic. Local remedies
are the most important. In *xeroderma*, any plan which systematically keeps the surface greased and slightly stimulated
will benefit. It is immaterial what grease is used. In the
horny forms of disease, a clean surface may be very readily
obtained by careful soaking with glycerine, by poulticing or
fermenting. The best plan is to use an alkaline bath, or a
warm alkaline lotion, to soften up the masses. After these
are removed in part by picking them away, the whole surface
can be greased and an alkaline bath used twice a week ($\mathfrak{z}$iv–
vi. of carbonate of soda and bran to the usual quantity of
of water). In this way the disease may be controlled so as
to prevent it being not only a disfigurement, but a discomfort,
save with occasional attention in winter.*

TUBERCULAR SKIN DISEASES.

LUPUS.

TILBURY FOX, M. D., LONDON, ETC.

In the majority of cases the *real* treatment of this affection
consists in the destruction of the lupoid tissue by caustics.
But general remedies are also needed. Lupus patients, especially the young, are often flabby, pale, anæmic, and, in a fair

*Skin Diseases, London, 1869, p. 291.

16

proportion of cases, phthisical. The use of cod-liver oil and iron in large doses is therefore indicated. Any weakness of digestion present is to be remedied first of all. Change of air should be secured if possible. A moderate amount of stimulants is beneficial.

Where the disease assumes the non-exedent form and the tubercles are well developed, and when the patient is not debilitated, a short course of bicyanide of mercury (gr. 1-16 to 1-12 for a dose) with bark is useful. In the ulcerating form, Donovan's solution will be found useful in connection with cod-liver oil. Constitutional remedies may be alone relied upon if the disease is not extending.

As regards local treatment, the erythematous variety of lupus wants ordinary stimulation with such an ointment as,

734.   R.   Olei juniperis pyrolignei,   f.ʒij.
             Adipis.   ʒj.   M.
       To be used every night.

Or,

735.   R.   Acidi carbolici fluidi,   f.ʒj.
             Glycerinæ,   f.ʒiss.   M.
       To be applied once a day, if it will be borne.

Whichever of these ointments is used it is to be followed with the application of dilute citrine ointment; or if there be much heat, by calamine powder, aside of zinc and glycerine and perhaps a little lead lotion. The application of collodion, when the disease is disappearing, helps in the cure very much.

If these measures do not appear to be successful, the use of *potassa fusa*, with an equal quantity of water, first to a limited part of the edge and gradually, at intervals of several days, to other parts, will, with general remedies, cure the disease. The *acid nitrate of mercury* is almost as useful. When the application of caustics is followed by much discomfort, heat, and swelling, then the stimulating plan is the best.

In the *non-exedent* form our author prefers the acid nitrate of mercury, or

736. ℞. Hydrarg. iodidi rubri,   gr.x-xx.
       Glycerinæ,   f.℥ss.   M.

Or,

             Equal parts of potassa fusa and water.

In the *exedent* form the solid silver caustic is the best. It must be deliberately and freely applied, chloroform being given, if necessary. The following is prefered by some.

737. ℞. Zinci chloridi,   ℥iv.
       Antimonii chloridi,   ℥ij.
       Amyli,   ℥j.
       Glycerinæ,   q. s.   M.

Others again commend nitric acid, mixed into a paste with sulphur, and laid on with a spatula. After caustic application a poultice should be applied and the surface dressed with a soothing ointment.

In all cases where the disease has been arrested and tends to heal, any mild stimulant or astringent application may be used, such as

GLYCERINUM ACIDI TANNICI.

738. ℞. Acidi tannici,   ℥j.
       Glycerinæ,   f.℥iv.   M.
       Rub together in a mortar, then transfer the mixture
       to a porcelain dish and apply a gentle heat until
       complete solution is effected.

Or,

739. ℞. Argenti nitratis,   gr.xx-xxx.
       Spir. ætheris nitrosi,   f.℥j.   M.

It must be remembered that local remedies act in efficiency in proportion to any improvement in the general health, brought about by internal remedies. The disease *can* be made much worse by caustics.

The following is a serviceable caustic :

740. ℞. Iodinii,   ℥ss.
           Potassii iodidi,   ℨj.
           Aquæ destillatæ,   f.℥v.   M.

741. ℞. Liquoris potassæ,   f.℥j
           Aquæ destillatæ,   f.℥j.   M.

A useful application.

742. ℞. Phosphori,   gr.x.
           Olei amygdalæ dulcis,   f.℥j.   M.
Dose.—Five to ten minims in emulsion.

W. H. GEDDINGS, M. D., OF AIKEN, S. C.

Lupus erythematojus is the most difficult of all skin diseases
to treat. The fact that the disease occasionally gets well of
itself, leaving only a flat thin cicatrix, admonishes us to be
careful in the selection of our remedies, and not to make use
of those caustics which produce thick and uneven scars. The
first step in treatment is, to remove the scales, for which purpose strips of linen, soaked in oil (any oil will answer) should
be applied to the diseased surface, over which a piece of flannel should be bound. This application is to remain on until
the scales become so soft that they can be rubbed off with
ease. After this preparatory treatment has been completed,
any of the numerous remedies which have been proposed in
the treatment of this disease may be applied. The simplest
treatment is that with potash soap (sapo virid.), which should
be applied as follows: Moisten a piece of flannel with lukewarm water, and lay a small quantity of soap upon it, and
then rub the diseased patch with it until a good lather is
formed, after which some of the soap should be spread upon a
piece of flannel and laid upon the diseased surface. The frictions are to be repeated daily, reapplying the flannel after
each operation. After the third day, the treatment should

be suspended and a new epidermis allowed to form. As soon
as this has taken place, the part should be well washed with
water, to determine whether the new epidermis is healthy.
If it stand the washing, the disease may be considered cured;
if not, the whole process is to be repeated. This simple treatment will, in some few cases, effect a cure; but in the majority
it will be found necessary to resort to more energetic remedies. Hebra has sometimes effected a cure by cauterizing
the part with strong liq. ammoniæ. It should be applied
with a brush made of picked lint. After each application
the diseased surface pours out a fluid not unlike that which
we see in a case of moist eczema. The ammonia should be
applied daily. Another application is a solution of iodine in
glycerine:

743.	R.	Iodinii,	Ɋj.	
		Potassii iodidi,	℥ss.	
		Glycerinæ,	f ℥j.	M.

It should be applied three or four times daily, until a thick
brown crust is formed. Its application should then be suspended until the crust falls off and enables us to see the condition of the skin underneath. Should it be necessary, the
caustics should be reapplied. The application of iodine and
glycerine is exceedingly painful.

Lac sulphur in the form of paste has been recommended.

744.	R.	Lac sulphuris,	℥ij.	
		Alcoholis,		
		Aquæ destillatæ,	aa. f.℥ij.	M.

Spread over the diseased surface.

After the crust has fallen off and the reaction has subsided,
the paste should be reapplied.

Should the above remedies fail, a strong solution of caustic
potash should be used:

745. ℞. Potassæ fusæ, ℥j.
Aquæ, f.℥ij. M.

The solution should be applied with a pencil made of picked lint. Immediately after the application, the part should be rubbed with cold water until a lather is produced. The latter procedure is necessary in order to prevent the potash from destroying the healthy as well as the diseased tissue. When the disease is located near the eye, that organ should be carefully closed and protected before any of the above applications are made.

Arsenic, our great sheet anchor in the treatment of chronic cutaneous affections, exerts no influence upon the course of lupus erythematosus. The same may be said of mercury, cod liver oil, and other internal remedies, all of which have been repeatedly tried without success.*

PARASITICAL DISEASES.

SCABIES.

TILBURY FOX, M. D., LONDON, ETC.

Scabies never gets well spontaneously. We must treat, 1st, the scabies itself, killing the acari and their ova; 2d, the secondary effects; and 3d, the complications. In all cases, *to all papules and vesicles*, the following should be applied:

746. ℞. Sulphuris, ℥ss.
Hydrargyri ammoniati, gr.iv.
Creasoti, gtt.iv.
Olei anthemidis, gtt.x.
Adipis, ℥j. M.

This is rubbed in night and morning; the same shirt kept on till the third day, when it is changed and a warm bath

*American Journal of the Med. Sciences. July, 1869, p. 65.

given; the ointment to be freely rubbed into the wrists and interdigits especially. In complicated scabies, we should treat the scabies always, scrupulously seeking out every suspicious papule; engrafting upon this the plan best suited to the complicated eruption, whatever it may be. In complicated scabies a small amount of acari may exist with a good deal of eruption. When the scabies itself in severe cases is well, a certain period must necessarily elapse before the secondary eruptions can be cured. The process of repair takes time. Hence the sulphur treatment must not be persisted in till all eruption has subsided in cases of severity. The case of scabies is to be judged of by the decrease and cessation of itching, and of the vesicles and papules.

If the sulphur treatment be pushed too far, an irritable erythematous state of the skin will be produced, which is often mistaken for the continuance of the disease. The cure is often retarded by the neglect of cleanliness, especially in regard to clothing. On the third day, when fresh linen is put on, it is best to destroy that taken off, or at any rate, to scald it thoroughly. In some cases the skin is too irritable to bear the sulphur; in that case, the unguentum potassii iodidi (which our author prefers in chronic scabies) is best. With fastidious folk, the following lotion may be used:

747. R. Hydrargyri chloridi corrosivi, gr.iij.
Aquæ, f.ℨvj. M.

Sulphur baths in the treatment of scabies our author never employs. They always seem to him to do harm. He has had cases under his care in which the irritation produced was excessive and troublesome to alleviate. If it be necessary to remove crusts, alkaline baths may be used. The following ointments are also of service:

748 ℞. Potassii sulphureti, ℥vj.
Saponis albi, lb.ij.
Olei olivæ, Oij.
Olei thymi, f.℥ij.

749. ℞. Potassæ sulphatis,
Sodæ sulphatis, aa. ℨxv.
Sulphuris præcipitati, ℨx.
Olei olivæ, f.℥ij. M.

750. ℞. Sulphuris iodidi,
Potassii iodidi, aa. ℨjss.
Aquæ, Oij. M.

751. ℞. Pulveris anthemidis, ℥ss.
Olei olivæ, f.℥ss.
Adipis, ℥ss. M.

This ointment is said by BAZIN to cure in three frictions.

PROF. S. D. GROSS.

752. ℞. Pulveris potassæ nitratis, ℈ij.
Sulphuris, ℨj.
Adipis, ℨj.¼.
Olei bergamii, gtt.v. M.

For scabies.

PROF. HARDY, OF PARIS.

753. ℞. Potassæ carbonatis, ℥ss.
Sulphuris sublimati, ℨj.
Adipis, ℥ij. M.

This is the ointment which our author employs in his speedy method, which he terms his two-hour cure.

He begins with a gentle friction with soft soap for the space of half an hour; secondly, a warm bath of half an hour's duration; and thirdly, a thorough inunction over the whole body with eleven ounces of his diluted ointment. The ointment remains in the skin for twelve hours (e. g., during the night), and is washed off in the morning in a tepid soap bath.

PROF. HEBRA, OF VIENNA.

754. R.   Sulphuris sublimati,
           Olei fagi seu cadini,   aa. ℥iij.
           Adipis,
           Saponis mollis,   aa. ℥viij.
           Cretæ praparatæ,   ℥ij.   M.

For scabies.

DR. H. B. SPENCER, OF OXFORD, ENG.

Our author calls attention to the great value of

UNGUENTUM POTASSII IODIDI

in the treatment of scabies.

It has been in his hands a cure *as certain* as sulphur ointment, and has the advantage that it does not smell or stain.
There is no need of the patient's going to bed, but it should
be well rubbed in all over the body night and morning, for a
few days.*

The following formulæ are useful in this affection:

UNGUENTUM SULPHURIS CUM HELLEBORA.

755. R.   Sulphuris sublimati,   ℥iv.
           Veratri radicis contritæ, ℥j.
           Potassæ nitratis,   ℥ss.
           Saponis mollis,   ℥iv.
           Adipis,   ℥xij.   M.

VEZINE'S FORMULA.

656. R.   Sulphuris sublimati,   ℥vj.
           Pulveris hellebori albi,   ℥ij.
           Potassæ nitratis,   gr.x.
           Saponis albi,
           Adipis,   aa. ℥vj.   M.

UNGUENTUM SULPHURIS CUM ANTHEMIDE.

757. R.   Unguenti anthemidis,   ℥vij.
           Sulphuris sublimati,   ℥j.
           Potassæ carbonatis,   ℥ss.   M.

This is a mild ointment for scabies and well adapted for persons of sensitive skin and for children.*

ADOLF'S FORMULA.

758. ℞. Sulphuris sublimati,
Pulv. baccarum junip.,   aa. ℥ij.
Pulv. baccarum laurini,
Adipis,   aa. ℥ij.   M.

UNGUENTUM SULPHURICI ACIDI.

DUNCAN'S FORMULA.

759. ℞. Acidi sulphurici,   f ℨss.
Adipis,   ℥j.   M.

UNGUENTUM SULPHURIS CUM SAPONE.

760. ℞. Sulphuris sublimati,   ℨxiv.
Saponis domestici,   ℥ij.
Adipis,   ℥viij.   M.

LINIMENTUM SULPHURIS CUM GLYCERINA.

BOURGNIGNON'S FORMULA.

761. ℞. Sulphuris sublimati,   ℥iij.
Potassæ carbonatis,   ℥j.
Glycerine,   f℥vj.
Pulv. tragacanthæ,   ℨj.
Olei lavandulæ,
Olei caryophylli,
Olei cinnamomi,   aa. ♏xx.   M.

SAPO SULPHURETI POTASSII.

IADELOT'S FORMULA.

762. ℞. Potassii sulphureti,   ℥iij.
Olei olivæ,   f.℥ij.
Saponis albi,   ℨxvj.
Olei thymi,   f.℥j.   M.
Fiat sapo.

*Wilson; Diseases of the Skin. 7th Am. ed., p. 867.

SAPO SULPHURIS SULPHATIS.

MOLLARD'S FORMULA.

763. R. Sulphuris præcipitate, ʒx.
Sodæ sulphatis,
Potassæ sulphatis, aa. ʒxv.
Olei olivæ, f.ʒijss. M.
Fiat sapo.

SAPO SULPHURIS ET AMMONIÆ.

NEUMAN'S FORMULA.

764. R. Sulphuris sublimati, ʒvj.
Ammoniæ muriatis, ʒj.
Saponis mollis, t.ʒij. M.

SAPO SULPHURIS SALINUS.

EMERY'S FORMULA.

765. R. Sulphuris sublimati, ʒviij.
Alcoholis, f.ʒj.
Aceti vini, f.ʒij.
Calcis hydrochloratis, ʒij.
Sodii chloridi, ʒss.
Saponis mollis, f.ʒj. M.

UNGUENTUM HELLEBORI ALBI.

ERASMUS WILSON.

766. R. Hellebori albi pulveris, ʒj.
Adipis, ʒiv.
Olei limonum, ℳxx. M.

UNGUENTUM STAPHISAGRIÆ.

BOURGUIGNON'S FORMULA.

767. R. Staphisagriæ sem. recent., ʒiij.
Adipis, ʒv. M.
Digest in a sand-bath for twenty-four hours, at a
temperature of 212°, and strain through a fine sieve.

Dr. ERASMUS WILSON prefers this formula to any other.*

LOTIO HELLEBORI ALBI.

768. R. Pulv. rad. hellebori albi, ʒj.
Aquæ, . f.ʒxxxij.
Boil down to f.ʒxvj., strain and add
Spiritûs vini rectificati, f.ʒij. M.

*Diseases of the Skin. 7th Am. Ed., p. 772.

SKIN HOSPITAL, LONDON.

769. R. Hydrarg. chloridi corros.   gr.xx.
      Pulveris indigo,   gr.iij.
      Adipis,   ӡj.   M.
Used in parasitic and tubercular affections.

770. R. Hydrarg. oxidi rubri,   gr.xij.
      Hydrarg. chloridi mitis,   gr.viij.
      Adipis,   ӡj.   M.

771. R. Hydrarg. bisulphureti,
      Hydrarg. oxidi rubri,   aa. gr.vj.
      Creasoti,   ♏ij.
      Adipis,   ӡj.   M.
Used in parasitic, squamous, papular, vesicular and ulcerous affections.

772. R. Hydrarg. iodidi rubri,   gr.x.
      Hydrarg. sulphatis flavæ,   gr.xxx.
      Adipis,   ӡj.   M.
Used in cachectic, tubercular and ulcerous affections.

773. R. Sulphuris sublimati,   gr.xxx.
      Hydrarg. ammoniati,
      Hydrarg. sulph. nigri., aa. gr.x.
      Olei olivæ,   f.ӡij.
      Creasoti,   ♏iv.
      Adipis,   ad. ӡix.   M.
Used in parasitic, papular and chronic vesicular affections.

LONDON HOSPITAL.

774. R. Sulphuris sublimatis,   ӡiv.
      Helleboris albi,   ӡj.
      Potassæ nitratis,   ӡss.
      Saponis mollis,   ӡiv.
      Adipis.   ӡij.   M.

CHARING-CROSS HOSPITAL, LONDON.

775. R. Sulphuris,   ƺij.
      Hydrargyri ammoniati,   gr.v.
      Creasoti,   ♏v.
      Olei olivæ,   f.ӡiij.
      Adipis,   ӡj.   M.

UNIVERSITY HOSPITAL, LONDON.

776. R.	Sulphuris sublimati,	ʒss.
	Hydrargyri ammoniati,	gr.x.
	Acidi carbolici cryst.,	gr.v.
	Olei olivæ,	f ʒij.
	Adipis,	ʒvj. M.

ST. MARY'S HOSPITAL, LONDON.

777. R.	Sulphuris sublimati,	ʒss.
	Hydrarg. sulph. rubri sub.,	
	Hydrarg. ammoniati, **aa.**	gr.x.
	Creasoti,	℞iv.
	Olei olivæ,	f.ʒij.
	Adipis,	ʒvj. M.

SKIN HOSPITAL, LONDON.

778. R.	Sulphuris iodidi,	
	Sulphuris præcipitati, aa.	gr.x.
	Olei amygdalæ amaræ,	℞iv.
	Adipis,	ʒj. M.

Used in parasitic, tubercular, and sebaceous affections.*

FAVUS.

TILBURY FOX, M. D., LONDON, ETC.

Internally, good food ; plenty of fat ; cod-liver oil and iron ; together with change of air and cleanliness must be prescribed.

Locally ; the hair should be cut short ; the crusts must be removed by soaking with

779. R.	Sodæ hyposulphitis,	ʒiv.
	Glycerinæ,	f.ʒij.
	Aquæ,	ad f.ʒvj. M.

Or, if preferred with

---

* Squire's Pharmacopœia of the London Hospitals. London, 1863, pp. 179, 180, 182.

780. R. Soda hyposulphitis, $\tilde{3}$iij.
       Acidi sulphurosi diluti, $\tilde{3}$ss.
       Aquæ,   ad Oj.   M.

When the scalp is cleaned, each hair must be extracted one by one, and parasiticides applied at once. One author prefers

781. R. Sodæ biboratis,   3j.
       Hydrargyri chlor. cor., gr.x—xx.
       Aquæ,   f.$\tilde{3}$ij—iij. M.

A certain portion of the surface should be cleaned each day, and the whole head meanwhile kept moistened with sulphurous acid lotion.

When our author wants to cure a favus case he epilates and applies his parasiticide himself. It takes time and is very troublesome. When the amount of parasite has been diminishing, as ascertained by the microscope, it is then advisable to exclude the air by the free use of unguents, after a good application of some parasiticide; the after-baldness must be remedied by stimulation.[*]

TINEA TONSURANS.

The treatment of *ringworm of the scalp* is a very tedious and difficult matter when the disease has lasted any time. In the earliest stage free blistering of each patch will suffice, with the free use of white precipitate ointment afterwards. One plan perseveringly followed is the best way to cure the disease.

In severe and more chronic forms cod-liver oil, quinine, good food, and change of air are often needed, and plenty of fat should be eaten with the food.

Locally, the object should be to get away all the fragments of the diseased hairs lodged in the follicles, and full of the

[*] Skin Diseases. London, 1869, p. 327.

spores. For this purpose apply a blister, if the hairs do not come away with any readiness. The healthy hair should be cut for a little distance around the circumference of the patches, so that the remedies may be applied freely to prevent extension of the disease. After removing as many of the hairs as possible, our author generally shaves the patches with a not over sharp razor, and in that way often drags out (without pain) a good many of the hairs left in ; but day by day attempts should be made at extraction. Meanwhile the surface may be blistered with corrosive sublimate solution :

782. R. Hydrarg. chlor. corrosivi, ℈ij.
　　　Acidi muriatici diluti, f.℥ss.
　　　Alcoholis, f.℥ss. M.

Grease may be applied to exclude air and prevent the dissemination of the spores, and the whole head washed night and morning, and well sopped in hyposulphite of soda lotion (F. 780). This plan must be closely pursued if success is to be early and complete.

In medium cases, after clipping the hair off very short, use *Coster's paste*, i. e. :

783. R. Iodinii, ℨij.
　　　Olei picis liq. decolorati, f.℥j. M.

This applied once or twice, at intervals of four or five days, effects a cure. As matters mend, it is only necessary to use some parasiticide ointment, the white precipitate, sulphur, or

784. R. Hydrarg. ammoniati, gr.vj.
　　　Hyd. nitris-oxidi levigati, gr.vj.
　　　Adipis, ℨj. M.

Or,

Borax (℈ij. ad. aquæ f.℥j.)

The latter our author employs freely with success.

No one should treat parasitic disease except under the guidance of microscopic examination.

The following (recommended by STARLIN) is useful:

785. R. Sulphuris præcipitati,   ʒij.
          Spiritûs camphoræ,   f.ʒss.
          Glycerinæ,   f.ʒss.
          Hydrargyri bisulphureti,   Ɔss.
          Pulveris amyli,   ʒij.
          Aquæ,   ad Oj.   M.

786. R. Sulphuris,
          Unguenti picis,   aa. ʒj.
          Unguenti hydrargyri,   ʒiij.
          Glycerinæ,   f.ʒss.   M.

Useful in all forms of *tinea*.

TINEA CIRCINATA.

TILBURY FOX, M. D., LONDON, ETC.

This affection (known also as *herpes circinatus* or *ringworm of the body*) may be treated effectively in the early stage by the application of any parasiticide, such as acetic acid; ink, even; strong borax solution; hyposulphite of soda lotion (F. 780); corrosive sublimate lotion (gr.ij. to aquæ f.ʒj) or white precipitate ointment. But in some instances the disease crops up here and there over different parts of the surface, and no sooner does one patch fade or go but others appear. This shows a condition of system favorable to the growth of the fungus, and this is altered by remedies specially adapted to the lymphatic temperament. The dilute acids and bitters are given, or even arsenic, iron, quinine, cod-liver oil, as the case may be. Alkaline baths are useful. There is no internal specific; local remedies act efficiently when the general health is satisfactory. The diagnosis once correctly made, the case should be easy. In *eczema marginatum* a solution of corrosive sublimate (gr.ij—f.ʒj.) is recommended by Dr. ANDERSON.

Where the disease is obstinate, blistering with ordinary vesicating fluid is often efficacious. Care must be taken that too much irritation is not produced.*

The following are useful:

VESICATING, VEGETABLE, PARASITICIDA.

787. R. Tincturæ iodinii compositæ, f.ʒj.
Iodinii,   gr.x.
Potassii iodidi,   gr.xv.   M.
Used in chronic stages of vegetable parasitic diseases.

788. R. Pulveris cantharidis,   ʒij.
Acidi pyro-acetici concen-
trati,   f.ʒviij.
Acidi tannici,   ʒj.   M.
Macerate for a week and strain.
Used in *tinea decalvans*.

789. R. Acidi carbolici fluidi,   f.ʒj.
Glycerinæ,   f.ʒss.   M.
Used in *tinea*.

SKIN HOSPITAL, LONDON.

790. R. Pulveris cantharidis,   ʒj.
Acidi tannici,   ʒss.
Acidi acetici,   f.ʒiv.   M.
Macerate seven days and strain.

This is the "causticum cantharidis compositum" of the Pharmacopœa of this hospital, and is used in parasitical, tubercular and sebaceous affections.†

MILDER PARASITICIDES (FOR ORDINARY USE.)

791. R. Potassii sulphureti,   ʒiij.
Saponis mollis,   f.ʒj.
Aquæ calcis,   f.ʒviij.
Alcoholis,   f.ʒij.   M.
Used in the various forms of *tinea, scabies, etc.*

---
*Skin Diseases. London, 1869, p. 340.
†Squire's Pharmacopœias of the London Hospitals. London, 1869, p. 5.
17

792. ℞. Hyd. chloridi corrosivi,   gr.ij–iv.
         Ammoniæ muriatis,   ʒss.
         Alcoholis,   f.ʒss.
         Aquæ rosæ,   ad f.ʒvi.   M.
In *tinea versicolor*, scabies, prurigo.

793. ℞. Acidi carbolici fluidi,   f.ʒij.
         Glycerinæ,   f.ʒj.
         Aquæ rosæ,   ad f.ʒviij.   M.
Use in *ringworm of the surface especially.*

Also, in the same affection.

794. ℞. Sodæ biboratis,   ʒij.
         Glycerinæ,   f.ʒj.
         Adipis,   ʒj.   M.

795. ℞. Hydrargyri sulphatis flavi, ʒss.
         Olei amygdalæ dulcis.
         Glycerinæ,   aa. f.ʒij.
         Adipis,   ʒij.   M.
Used in *tinea.*

796. ℞. Unguenti hydrargyri nitra-
               tis,   ʒss.
         Sulphuris,   ʒij.
         Creasoti,   ℳx.
         Adipis,   ʒj–ij.   M.
Use in ordinary *ringworm* and *tinea sycosis.*

797. ℞. Cupri carbonatis,   ʒij.
         Adipis,   ʒj.   M.
Use in parasitic disease generally, especially *tinea sycosis.*

798. ℞. Calcis vivi,   ʒj.
         Sulphuris sublimati,   ʒv.
         Aquæ,   f.ʒxx.   M.
    Boil for half an hour and filter, making the quantity
       of fluid product ten ounces.
For *scabies, alphos, etc.*

799. ℞. Sulphuris iodidi,   gr.x–xx.
         Unguenti benzoati,   ʒj.   M.
For *sycosis, lupus erythematosus, etc.*

800. ℞ Sulphuris sublimati,
Zinci sulphatis, aa. ℥j.
Adipis præparati, ℥viij. M.

For *scabies.*

801. ℞. Sulphuris sublimati, ℥ss.
Zinci sulphatis, ℥iss.
Pulv. hellebori albi, ℨj.
Saponis mollis, f℥xiij.
Adipis, ℥ij.
Olei carui essentialis, f.℥ss. M.

For *scabies.*

SKIN HOSPITAL, LONDON.

802. ℞. Creasoti, ♏vj.
Hyd. oxidi rubri, gr.x.
Unguenti hydrargyri, ℥ss.
Adipis, ℥j. M.

Use in parasitic and cachectic affections.*

SKIN HOSPITAL, LONDON.

803. ℞. Sodæ hyposulphitis, gr.ix.
Acidi sulphurosi diluti, ♏xij.
Aquæ, f.℥j. M.

Used, diluted with 1 to 3 parts of water, in parasitic and squamous affections.

804. ℞. Sulphuris præcipitati, gr.vj.
Glycerinæ, ♏xij.
Hyd. sulph. rubri sublim. gr.ss.
Spiritûs camphoræ, ♏iss.
Amyli, gr.vj.
Aquæ, ad f.℥j. M.

Used, diluted with 1 to 3 parts of water, in parasitic, sebaceous and pustular affections.†

DR. H. S. PURDON, LONDON.

Parasiticides may be divided into those derived from the vegetable, animal, and mineral kingdoms ; but without going deeply into the subject, it may be briefly stated that the most

*Squire's Pharmacopœias of the London Hospitals. London, 1869, p. 175.
†Squire's Pharm. of the London Hospitals. London, 1869, p. 60.

valuable obtained from the first are iodine, creasote, carbolic
acid, and acetic acid. The last three check the development
of spores ; creasote, according to M. BEAUCHAMP, although it
allows the mycelium to form, prevents the spores from germi-
nating. From the second the only remedy in use is cantharides, which, when used in the form of the liniment of the
British Pharmacopœia (about the strength of the cantharidal
collodion, U. S. P.), quickly cuts short the disease, especially,
tinea tonsurans, circinata, and alopecia acuta ; it likewise
stimulates the affected skin to take on a more healthy action.
From the mineral kingdom we have mercury, especially the
bichloride, chromate, nitrate, and white precipitate, sulphur,
borax, etc. The first has a well earned reputation, and the
chromate of mercury, our author is at present trying in tinea
versicolor, and some other forms of vegetable parasitic diseases ;
an objection to its use is that it does not mix with water,—
indeed, it is insoluble in any fluid, but may be used as an
ointment. He has added glycerine and rectified spirit, so as
to endeavor to suspend it in solution, but without success.
The only way to manage is to shake the bottle before applying it. A useful auxiliary to the above remedies is epilation,
which should be performed in inveterate cases. Of course,
constitutional treatment is of the utmost importance, quinine
being our chief remedy ; which substance, it is asserted, has
the property of destroying vegetable growth. Dr. BINZ* has
found that a neutral solution of chlorohydrate of quinine,
soluble in sixty times its volume of water, has the power of
destroying infusoria and fungi developed in vegetable infu-
sions, and believes it has a peculiar antiseptic action different
from other vegetable alkaloids. The tincture is the best
preparation for children.

*Medical Press and Circular, April, 1869.

No doubt the growth and development of a fungus is favored by some peculiar condition of the system; for example, tinea versicolor flourishes and is common on the bodies of consumptive patients.

In all cases of vegetable parasitic diseases, our author prescribes constitutional as well as local treatment. Cod-liver oil, pancreatinine, the syrup of the iodide of iron, quinine, and in hospital practice, salicine, are the remedies relied on. The therapeutical fact should be remembered that parasitical affections are rarely, if ever, "cured" by destroying the parasite; but they can be eradicated by administering appropriate tonics and alteratives which are capable of correcting the blood dyscrasia, which tends to keep up the disease.*

J. M. DA COSTA, M. D., PHILADELPHIA.

805. R. Calcis hyposulphitis,
Sodæ hyposulphitis, aa. ℥ss.
Aquæ, f.℥iv. M.

A useful lotion for sycosis mentis.

VIII. VENEREAL DISEASES.

GONORRHŒA AND GLEET.

WM. ACTON, M. R. C. S., ETC., LONDON.

806. R. Zinci. sulphatis,
Acidi tannici, aa. gr.ij.
Aquæ, f.℥ij. M.

To be used repeatedly during the day as an injection.

*Periscope of *Medical and Surgical Reporter* for August, 1860).

D. HAYES AGNEW, M. D., PHILADELPHIA.

807. ℞. Tincturæ cubebæ,   f.℥ij.
          Copaibæ,   f.℥j.
          Liquoris potassæ,   ℳ lxxx.
          Liq. morphiæ sulphatis,   f.℥j.
          Aquæ cinnamomi,
          Misturæ camphoræ,   aa. f.℥ij.
          Pulveris acaciæ,
          Sacchari albi,   aa. ℨij.   M
A tablespoonful three or four times a day.

FREEMAN J. BUMSTEAD, M. D., NEW YORK.

808. ℞. Argenti nitratis,   gr.1-6—1-4.
          Aquæ destillatæ,   f.℥j.   M.

Recommended as an efficacious and perfectly safe injection for the abortive treatment of gonorrhœa. It is to be used every two or three hours until the desired amount of substitutive inflammation is attained. It is adapted only to the commencement stage; if employed after the discharge has become purulent and pain is felt in passing water, it is almost sure to fail and delay the cure. This injection should only be used under the supervision of the surgeon.

M. A. CULLERIER, SURGEON TO THE HÔPITAL DU MIDI, ETC., PARIS.

809. ℞. Copaibæ,   f.ʒv.
          Cubebæ,   ʒiv.
          Spts. menthæ piperitæ,   q.s.   M.
       Fiat confectio.
From four to five drachms a day are given.

This formula is one of the most frequently employed at the Hôpital du Midi.

For the *abortive* treatment of gonorrhœa, our author uses large doses of Copaiba (f.ʒiv-v. a day) or cubebs (ʒv.–viij. a day). He considers them much more valuable than any of the abortive injections. They are to be employed only, however, when the gonorrhœa is of recent date, when there is little or no pain, and where the discharge is not as yet muco-prurulent.

Under favorable circumstances, when the abortive treatment is thus employed, the discharge will diminish, or disappear in the course of four or five days. The treatment should not then be suspended, but, on the contrary, continued for several days after the cure is apparently complete. If this precaution be neglected, the inflammation may re-appear. If, after from six to eight days, no improvement is manifest, it is useless to persist longer in this form of treatment. Astringent injections should not be combined with this use of the balsam. They have no advantage at this early period of the disease, and often keep up an amount of irritation, which may interfere with the effect of the internal remedy.

When the inflammatory period of gonorrhœa is over; when the discharge is unaccompanied with pain and erections have ceased ; when, in short, there is no longer any danger of excessive irritation from local treatment, Cullerier advises injections to complete the cure.

(Dr. C. P. JUDKINS, of Cincinnati, gives in the Cincinnati *Lancet and Observer*, for March, 1869, a number of cases of gonorrhœa treated, in their acute stage, in the manner recommended above by Cullerier. His results were very favorable.)

The following injections are those most frequently prescribed at the Hôpital du Midi :

810. R. Zinci sulphatis,
Plumbi subacetatis. aa. gr.xv.
Aquae, f.℥iv. M.

811. R. Aluminis, ℥iss.
Aquae, f.℥iv. M.

812. R. Acidi tannici,   gr.vij.
          Aquæ,   f.ʒj   M.
Two injections a day are sufficient. Before each injection
the patient should urinate.

813. R. Bismuthi subnitratis,   ʒv-ʒj
          Aquæ,   f.ʒvj.   M.

This injection is useful in chronic discharges. The almost
insoluble salt is deposited on the walls of the urethra and separates them, producing its good effects rather by this isolation
of the surfaces than by its medicinal action. The patient
should prepare this injection himself when he uses it. If kept
a day or two it becomes acid and does harm.

The subacetate of lead is sometimes prescribed alone as
follows:

814. R. Plumbi subacetatis,   gr.xv.
          Aquæ,   f.ʒiij.   M*.

P. DIDAY, EX-CHIRURGIEN EN CHEF DE L'HOSPICE DE
L'ANTIQUAILLE, DE LYON, ETC.

Our author gives the following precepts on abortive medication:

1st. The physician should inform all his patients of the
possibility of a speedy cure if they will consult him in time.
He should indicate to them the first signs of the disease and
emphasize the danger of any delay.

2d. The patient having come the physician should operate
immediately and operate himself.

3d. One injection suffices, that is to say, one sitting. Two
injections should be given, one to clean the canal from urine
and muco-pus, and then the second to act curatively.

4th. The following formula should be used:

815. R. Argenti nitratis crystallazi, gr.iv.
          Aquæ destillatæ,   f.ʒiv.   M!

*Atlas of Venereal Diseases, pp. 102-107.

Only about a drachm and a half of this liquid should be drawn into the syringe. For if consultation is had in time, the inflammation has not extended, and cauterization to the depth of two inches of the canal suffices.

6th. After the cleansing injection, the second, that which acts, ought to be retained three minutes, and ought to distend the anterior part of the urethra.

7th. Care should be taken that the *fossa navicularis*, the starting point of the disease, is well bathed with the injection.*

SILAS DURKEE, M. D., ETC., BOSTON.

816. R. Copaibæ, f.℥iij.
 Spiritûs ætheris nitrosi,
 Tincturæ kino, aa f.℥ss.
 Morphiæ sulphatis, gr.iv.
 Aquæ camphoræ, f.℥ij. M.
 One teaspoonful ter die.

Usually, an efficient check will be put to the gonorrhœa in eight or ten days by the use of this preparation.†

A combination of the balsam with the powder of cubebs makes a good compound, and may be employed according to the following formula :

817. R. Copaibæ, f.℥j.
 Pulveris cubebæ, ℥j.
 Liquoris potassæ, f.℥iij.
 Morphiæ sulphatis, gr.ij.
 Aquæ camphoræ, f.℥iv.
 Aquæ cinnamomi, f.℥ij. M.
 A dessertspoonful ter die.

In whatever form the balsam is administered, it should be continued in gradually diminishing quantities for ten or twelve days after the discharge has entirely ceased.‡

---

*Nouvelles Doctrines sur la syphilie, p. 90.
†A Treatise on Gonorrhœa and Syphilis. 5th Ed , p. 38
‡A Treatise on Gonorrhœa and Syphilis, 5th Ed., p. 41.

Capsules of capaiba may be taken when the liquid balsam disagrees. They should be administered soon after eating, and as freely as the stomach will bear. Capsules containing a combination of copaiba and the extract or oil of cubebs are sometimes more efficient than those composed of *copaiba alone*.

In cases in which it is important to combine with the co- paiba remedies that exert an anodyne influence upon the organs, as well as a modifying agency upon the qualities of the urine, our author recommends the following formula of WILLIAM ACTON, of London, for an

ELECTUARY.

818. R. Copaibæ   ʒvj.
          Magnesiæ,   ʒiss.
          Extracti hyoscyami,   ʒss.
          Pulveris camphoræ,   ʒj.
          Theriacæ,   ʒiij.
          Micæ panis,   ʒjss.   M.

For an electuary. Dose—one drachm ter die.

This is a favorite prescription of Acton, who claims that the magnesia neutralizes the urine, that the hyoscyamus allays irritation of the bladder or prostate, and that the camphor checks any disposition to involuntary erections, which without it often become a troublesome complication.

COPAIBA AS AN INJECTION.

Our author gives the formula of Dr. DICK, of London, for the oil of copaiba employed as an injection :

819. R. Olei copaibæ,   f.ʒj.
          Pulveris acaciæ,   ʒij
          Aquæ,   f.ʒvj.   M.

In subacute gonorrhœa and in gleet this injection is to be used twice a day for a few days ; afterward more frequently.

The formula of VELPEAU is as follows :

820. ℞. Copaibæ,			f.ʒij.
       Tincturæ opii,		f.ʒss.
       Mucilaginis acaciæ,	f.ʒiss.			M.

For an injection, to be repeated twice or thrice a day.

It is asserted that successful results have been obtained in this manner in cases in which the balsam could not be tolerated by the stomach.*

CUBEBS.

Our author has employed the annexed formula for a number of years.

821. ℞. Pulveris cubebe,		ʒviij.
       Pulveris alumnis,		ʒj.
       Pulveris cinnamomi,	ʒj.			M.

For xxxij. powders. One ter die.

This combination of cubebs and alum will usually diminish the urethral discharge in two or three days, and if the patient will observe a perfectly quiet state of the body, he will find that in eight or ten days the gonorrhœa will be nearly at an end. The strictest avoidance of exercise constitutes an important element in the treatment of every case of gonorrhœa, and the patient should even keep in a recumbent posture in order to secure the best effects in the shortest time.

As with the balsam copaiba, so with cubebs; they should not be discontinued under a fortnight after the cessation of the urethral discharge.†

The tincture is an elegant and convenient form of administering cubebs. It may be given in doses of f.ʒj. to f.ʒij. four or five times a day, or combined thus:

822. ℞. Tinct. cubebæ,		f.ʒij.
       Tinct. cantharidis,	f.ʒiss.
       Morphiæ sulphatis,	gr.ij.
       Aquæ camphoræ,		f.ʒiij.			M.

A dessertspoonful ter die, in half a gill of cold water.

*A Treatise on Gonorrhœa and Syphilis. Fifth Ed., p. 52.
†Treatise on Gonorrhœa and Syphilis. Fifth Ed., p. 43.

Or, the fluid extract may be used in this manner :

823.　℞.　Ext. cubebæ fluidi,　　　　f.ℨiv.
　　　　　Morphiæ sulphatis,　　　　gr.ij.
　　　　　Mucilaginis acaciæ,
　　　　　Aquæ camphoræ,　　aa. f.ℨij.　　　M.

Our author also recommends the following formulæ of Drs.
Druitt, Langston Parker, Beyran and Holmes Coote:

824.　℞.　Copaibæ,　　　　　　　f.ℨss.
　　　　　Olei Cubebæ,　　　　　　f.ℨss.
　　　　　Liquoris potassæ,　　　　f.ℨiij.
　　　　　Spiritûs myristicæ,　　　f.ℨss.
　　　　　Aquæ camphoræ,　　　　ℨj.　　　M.
　　　　Two tablespoonfuls ter die.

The combination of copaiba with the oil of cubebs, as
above, will sometimes be found to agree better with the stomach than the capsules or any other combination.

In chronic gonorrhœa or gleet the balsam and the cubebs
may be advantageously combined with iron, as follows:

825.　℞.　Pulveris cubebæ,　　　　ℨss.
　　　　　Copaibæ,　　　　　　　　f.ℨij.
　　　　　Ferri sulphatis,　　　　　ℨj.
　　　　　Terebinthinæ chiæ,　　　ℨiij.　　M.
　　　　To be made into boluses of gr.x each.
　　　Dose—From 15 to 30 a day. Usefully employed in lax
　　　constitutions.

826.　℞.　Pulveris cubebæ,　　　　ℨj-ij.
　　　　　Ferri carbonatis.　　　　ℨss-j.　　M.
　　　For one powder, to be taken ter die.

Particularly useful after the acute symptoms have subsided.

ELECTUARY.

827　℞.　Copaibæ,　　　　　　　f.ℨiss.
　　　　　Magnesiæ,　　　　　　　ℨj.
　　　　　Pulveris aluminis,　　　gr.xv.
　　　　　Pulveris catechu,　　　　ℨiss.
　　　　　Pulveris opii,　　　　　gr.xv.
　　　　　Spiritûs menthæ piperitæ,
　　　　　Spiritûs canellæ,　　aa. gtt.xl.　　　M.

For an electuary, in subacute gonorrhœa at the commencement of the discharge and in gleet. A teaspoonful ter
die in a moistened wafer.

828. ℞. Copaibæ, gtt.xv.
Cubebæ, ʒj.
Spiritûs ætheris nitrosi, gtt.xx.
Misturæ acaciæ, f.ʒj.
Aquæ camphoræ, f.ʒx. M.

This is the preparation used at Bartholomew's Hospital. The above quantity is to be taken ter die.

ASTRINGENT INJECTIONS.

829. ℞. Plumbi acetatis,
Zinci sulphatis, aa. gr.iij.
Aquæ rosæ, f.ʒvj. M.
To be used ter die.

830. ℞. Zinci sulphatis,
Plumbi acetatis, aa. gr.xv.
Tincturæ catechu,
Tincturæ opii, aa. f.ʒj.
Aquæ rosæ, f.ʒvj. M.

This formula is recommended by Ricord and Acton, as well as by our author.

For *painful* erections the following formula is very valuable:

831. ℞. Extracti belladonnæ, Əj.
Lupulinæ recentis,
Pulveris camphoræ, aa. ʒj. M.
For 48 pills. One to four at night.

Or,

832. ℞. Pulveris camphoræ,
Extracti lactucæ, aa. Əij. M.
For xx pills. One to six at night.

For *chordee* the following surpasses all other remedies:

833. ℞. Spiritûs camphoræ, f.ʒj.
For one dose, to be taken in sweetened milk, on going to bed. Every time the patient wakes with the chordee he is to rise and repeat the dose.

In the treatment of

GLEET,

our author employs the following :

834. ℞. Tincturæ cantharidis,
Olei terebinthinæ,   aa. f.℥j.
Mucilaginis acaciæ,   f.℥ij.   M.

A tea-spoonful ter die, together with the following injection :

835. ℞. Acidi tannici,   ℈j.
Plumbi acetatis,   gr.viij.
Aquæ,   f.℥viij.   M.

A syringeful to be injected three or four times in the twenty-four hours.

The whole of the perineal integument in obstinate cases should likewise be made perfectly raw with the compound tincture of iodine, and the patient confined to his room.

BLISTERS IN GLEET.

These stand at the head of all local remedies in cases not dependent on stricture or otherwise complicated. The *cantharidal collodion*, which is preferable to the cerate, should be applied by means of a brush along the whole length of the penis, except two or three lines toward the preputial orifice. After the evaporation of the ether, the parts are to be protected by a linen rag. A second application is seldom required. Saturday evening should be chosen for the operation, so that the patient may remain at rest the following day.

Or, large blisters 3x3 may be applied high up on the inner surface of the thighs. They should be put on at bed time and allowed to remain until morning. The vesicated surface may be dressed with cold cream, the benzoated oxide of zinc ointment or lint soaked in castor oil. Very likely the urethral discharge and scalding on micturition may be increased for a day or two, for which reason the patient should be provided with the following antidote :

836. ℞. Spiritûs ætheris nitrosi, f.ʒj.
     Aquæ camphoræ, f.ʒij.
     Mucilaginis acaciæ, f.ʒss.   M.

Two teaspoonfuls every hour until four or five doses are taken, if necessary. Direct also a warm bath, if the symptoms be urgent, of which there is little probability.

INJECTIONS IN GLEET

837. ℞. Acidi nitrici, gtt.xx.
     Aquæ, f ʒviij.   M.

One fluid drachm to be injected every hour or even oftener if the patient choose.

This is one of the best as well as the cleanest of injections. It is tonic and astringent to the mucous surface. Or, the following may be used :

838. ℞. Strychniæ, gr.iv.
     Acidi nitrici, gtt.viij.
     Aquæ, f.ʒiv.   M.

Inject f.ʒj. ter die, after micturition. Order at the same time internally :

839. ℞. Extracti nucis vomicæ, gr.xij.
     Quiniæ sulphatis,
     Extracti hyoscyami, aa. gr.xxiv.   M.

For xxiv pills. Take two one hour before each meal.

The chloride of zinc makes a valuable injection in some cases of gleet and also in gonorrhœa, employed according to the following formula :

840. ℞. Zinci chlori, gr.vj.
     Aquæ rosæ, f.ʒvj.   M.

To be used twice in the twenty-four hours for a few days, after which more frequently, if no unpleasant effect is complained of by the patient.

It is hardly worth while to continue an injection if it does not exert the desired action in the course of one week. The mucous membrane requires some other kind of stimulant.

The instrument should be large enough to distend the urethra moderately and nothing more. An unmedicated bougie should be introduced at first, to the distance of five or six inches, and allowed to remain ten or fifteen minutes, according to the amount of irritation it excites. In some cases it may be repeated two or three times a day, in others not oftener than once in two or three days. If after a few trials with this simple bougie no effect is perceptible, it may be coated with some slightly stimulating ointment, thus:

841. ℞. Extracti belladonnæ,   ℥j.
           Unguenti hydrargyri,   ℥ss.   M.

The bougie should be introduced cautiously and withdrawn with a slight rotary movement, so as to cause an equal distribution of the ointment over the whole surface of the canal. This medicated instrument gives considerable pain and excites inflammation, so that usually an increase of the discharge is occasioned for a few days, after which it often entirely ceases. If this does not happen, a second trial should be made in ten or twelve days. The following substances may also be used for lubricating the bougie:

842. ℞. Argenti nitratis,   gr.j.
           Cerati simplicis,   ℥j.   M.
Or,

843. ℞. Ung. hydrarg. nitratis,   ℥j.
           Olei olivæ,   f.℥ij.   M.
Or,

844. ℞. Olei terebinthinæ,   f.℥ss.

Constitutional treatment of gleet. Internal remedies alone seldom result in the cure of this affection. Tonics are generally useful. A plain but substantial diet should be allowed, of which lean meats should constitute a part. The subjoined formulæ are the best which can be used. If they fail, it is hardly worth while to waste any more time upon internal remedies, but blistering should be resorted to as before recommended (p. 270).

845. ℞. Copaibæ,   f.℥ss.
     Tincturæ cantharidis,   f.℥ij.
     Tinct. ferri chloridi,   f.℥j.   M.

xxx gtt. ter die, in a gill of sweetened water.

846. ℞. Tincturæ cantharidis,   f.℥j.
     Quinæ sulphatis,   ℥ss.
     Acidi sulphurici diluti,   f.℥ss.
     Tincturæ ferri chloridi,   f.℥ij.
     Aquæ,   f.℥viij.   M.

Two tablespoonfuls ter die, in an equal quantity of cold water.

When there is a thickened and uneven condition of the urethra, the following is appropriate:

847. ℞. Hydrarg. iodidi rubri,   gr.iij.
     Potassæ iodidi,   ℥iss.
     Alcoholis,   f.℥ss.
     Syrupi aurantii,   f.℥ijss.   M.

xxx gtt. ter die.

The combination of iodide of potassium with iodine is useful, especially in those who show a scrofulous diathesis:

848. ℞. Potassii iodidi,   ℥iij.
     Iodinii,   gr.j.
     Aquæ destillatæ,   f.℥vij.   M.

A teaspoonful ter die.

849. ℞. Pulveris ergotæ,   ℥iss.
     Ferri carbonatis,   ℥ij.
     Pulveris vanillæ,
     Pulveris camphoræ,   aa. gr.vj.   M.

For xxiv powders. One morning and evening.

850. ℞. Olei terebinthinæ,   f.℥ij.
     Morphiæ sulphatis,   gr.j.
     Aquæ camphoræ,
     Mucilaginis acaciæ,   aa. f.℥ij.
     Sacchari albi,   ℥iij.   M.

A dessertspoonful ter die.*

---

*A Treatise on Gonorrhœa and Syphillis, 5th Ed., p, 60, et seq.

18

TREATMENT OF BALANITIS.

This affection, when uncomplicated, is quickly cured. After the parts are exposed by drawing back the prepuce and cleansed by bathing in tepid water, the best topical application for the slight abrasions and small patches of aphthæ is the following:

851. R. Liquoris sodæ chlorinatæ,   f.℥ss.
         Aquæ,   f.℥vij.   M.

This solution is to be applied on pieces of lint between the prepuce and the glans, three or four times a day.

If the erosion be considerable and the puriform exudation copious, an astringent lotion may be appropriate, thus:

852. R. Zinci sulphatis,   gr.ij.
         Acidi tannici,   gr.iv.
         Glycerinæ,   f.℥ij.
         Aquæ,   f.℥iv.   M.
      Apply with lint.

Simple *lime water* will frequently effect a rapid cure.*

GONORRHŒA IN THE FEMALE.

853. R. Tincturæ lyttæ,
         Tincturæ cubebæ,   aa. f.℥ij.
         Morphæ sulphatis,   gr.iv.
         Aqua camphoræ,   f.℥iv.   M.
      A teaspoonful ter die, in a gill of rice water, or toast water in cases of *vaginal gonorrhœa*.

The following is an excellent tonic and astringent injection, after the more acute symptoms have subsided:

854. R. Acidi nitrici,   gtt.xx.
         Infusi cinchonæ rubræ,   f.℥x   M.
      To be used repeatedly during the day.

*A Treatise on Gonorrhœa and Syphilis, 5th Ed. p 75, et seq.

The subjoined injection is also appropriate:

856. ℞. Acidi tannici, ℈ij.
Potassæ chloratis, ℥j.
Aquæ, f.℥xvj. M.

The patient should use an ordinary syringeful at a time, and repeat the operation *ad libitum.* Its beneficial effects generally are apparent in a few days.*

PROF. S. D. GROSS.

856. ℞. Copaibæ, f.℥ss.
Spts. ætheris nitrosi, f.℥ij.
Tincturiæ opii, f.℥j.
Sodæ bicarbonatis, ℥j.
Pulveris acaciæ,
Sacchari albi, aa. ℥j.
Aquæ camphoræ, f.℥iv. M.

Tablespoonful three times a day.

PROF. WILLIAM A. HAMMOND, NEW YORK.

Our author only recognizes two forms of specific gonorrhœa; the one proceeding from the action of an indurated chancre on a mucous surface, the other from the action of the pus of a soft chancre on such a surface. The former he denominates syphilitic gonorrhœa. The gonorrhœa thus contracted from chancrous matter may communicate the disease, though in a less virulent form, to others and so on.†

Holding these views, Dr. H. prefers to conduct the treatment of *syphilitic gonorrhœa* altogether by injections. Copaiba and cubebs do not appear to have much effect upon it, though occasionally during the first day or two they may lessen the discharge from the urethra. In women they cannot possibly exert the slightest influence. Neither has he any predilection for the abortive plan of treatment with strong injections.

857. ℞. Argenti nitratis, gr.¼-½.
Aquæ destillatæ, f.℥j. M.

This injection will be found very efficacious, repeated six or

*A Treatise on Gonorrhœa and Syphilis, 5th Ed., p. 173.

seven times in the twenty-four hours, the patient retaining the
fluid in contact with the diseased urethra for at least a minute.   He should press the penis between the thumb and fingers of the left hand, so as to close the urethra, and thus preventing the injection passing into the bladder, or he should
sit on the edge of a chair so that the urethra is constricted in
the perineum.   The meatus should be closed with the right
hand as the syringe is withdrawn.

Another excellent injection is:

   858.   R.   Zinci chloridi,   gr.$\frac{1}{4}$–ij.
               Aquæ,   f.ȝj.   M.

Tannin, sulphate of zinc, acetate of lead and alum may be
used as injections in the proportions of gr.ij–iij to f.ȝj. of
water.

Permanganate of potassa possesses advantages over other
substances used as injections in gonorrhœa.   It seems to exercise a certain amount of specific influence by destroying the
contagious property of the secretion from the mucous membrane.

   859.   R.   Potassæ permanganatis, gr.$\frac{1}{4}$–ij.
               Aquæ,   f.ȝj.   M.
      The weaker solution should be used first, and gradually in-
        creased.   Eight or ten injections should be made in the
        twenty-four hours.

While the treatment is going on, it is advantageous to
render the urine less acrid and thus to lessen the painful micturition.   For this purpose demulcent drinks, such as flaxseed
and slippery elm teas, either acidulated or not, with lemon
juice, tartaric or citric acid, or the bitartrate of potash, may
be freely drank.   If there is any very great amount of pain,
opiates may be given with or without camphor, but they are
rarely necessary.

Occasionally benefit is derived from the administration of

saline cathartics, but it is not advisable to continue them as regular means of cure.

For chordee, camphor should be given at night in the dose gr. iv–v. Cold water douches to the penis are, however, more efficacious.

The diet should be unstimulating, but not low; sexual intercourse should be avoided; moderate exercise should be indulged in, and the life should be regular and temperate in all things. Hot or cold baths, as the patient may prefer, should be taken daily.

The treatment of syphilitic gonorrhœa in women is usually even easier than in men. The above injections may be used, the strength, however, being doubled. They should be frequently administered, the patient being in the recumbent posture, with the hips elevated, so as to retain the fluid in contact with the parts for a minute or two. Frequent injections of cold water in the intervals will add much to the comfort and facilitate the cure.

In the management of syphilitic gonorrhœa, great care should be taken to avoid causing a solution of continuity in any part of the tissue over which the discharge passes, or with which it may come in contact, or a chancre may be the consequence.

In the treatment of *simple gonorrhœa* (*i. e.*, that caused by the virus of a non-infecting chancre, or by the discharge produced in others by such a chancre), after the discharge is well established, reliance should be placed upon injections. Those recommended in syphilitic gonorrhœa will be found most advantageous.

The following mixture of copaiba is capable of doing more good than the uncombined balsam, and it is not much more disagreeable to the taste or stomach:

860. R. Copaiba, f.ʒij.
Spts. ætheris nitrosi, f.ʒj.
Tincturæ opii,
Tincturæ iodinii, aa. f.ʒj.
Magnesiæ, ʒij.
Mucil. acaciæ, f.ʒv. M.

One to two teaspoonfuls ter die.

No internal treatment should be depended upon to the exclusion of injections.

Stimulants should be avoided, as should also *salt meat.*

In the management of the chronic stage of simple gonorrhœa or *gleet,* the affected individual should be placed upon a good, plain, nutritious diet, and the mind and body pleasantly and systematically employed. The greatest benefit is derived from cold plunge baths, followed by frictions of the skin with coarse towels or hair brushes. As internal remedies use

861. R. Ferri sulphatis, gr.ij.
Quiniæ sulphatis, gr.½. M.

For one pill ter die.

The oxalate or citrate of iron may be substituted in the same dose. In addition, our author has derived great benefit from the use of the following recipe:

862. R. Tincturæ cantharidis, f.ʒss.
Strychniæ, gr.j.
Syrupi limonis, f.ʒiij. M.

A teaspoonful morning and evening.

Injections should be persevered with, changing one for another as they seem to lose their effect. Medicated bougies are valuable adjuncts. Mercurial ointments, carbonate of zinc ointment, and particularly iodine ointment prove useful.*

---

* Lectures on Venereal Disease. pp. 271, 272, 273, 281.

BERKELEY HILL, M. B., LONDON, F. R. C. S., ETC.

863. R. Potassæ bicarbonatis, ʒj.
Potassæ nitratis, ꝯj.
Ætheris, f.ʒss.
Tincturæ opii, ℳxxv.
Aquæ camphoræ, f.ʒj.   M.

Two tablespoonfuls every six hours in the *acute* or highly inflammatory stage, to render the urine less acid.

In this early stage, copaiba and cubebs are not beneficial, and only two injections of any service, viz., half hourly injections of tepid water, or hourly injections of alum or sulphate of zinc, gr. ¼ to f.ʒj. aquæ. The former are often useless, and the latter, if they increase the irritation, are to be stopped.

SUPPOSITORY FOR CHORDEE.

864. R. Morphiæ sulphatis, gr.⅛–¼.
Butyri cocoæ, gr.x.   M.

To be passed into the rectum on going to bed.

When the pain is violent 30 to 40 gtt. of tinctura opii in a wineglassful of decoction of starch should be injected.

Our author has repeatedly found of service in chronic gonorrhœa the following capsule devised by Sir Henry Thompson:

865. R. Ext. cubebæ ætherialis,
Olei copaibæ, aa. ℳiv.
Picis liquidæ, ℳij.   M.

For one capsule. One three or four times a day.

A very useful formula for injection is that of the "Four Sulphates:"

866. R. Zinci sulphatis,
Ferri sulphatis,
Cupri sulphatis,
Aluminis, aa. gr.x.
Aquæ, f.ʒviij.   M.

The solution is not used in full strength at first, but the first day is diluted with three times its bulk of water. If severe smarting follow, it is further diluted. Its strength is gradually increased until its full strength is used or the discharge stops. This being attained, it is diminished in strength step by step until plain water is reached. In this plan, ten days should be employed, and a pause allowed before any other treatment is adopted, should that prove necessary.*

DR. J. D. HILL, ROYAL FREE HOSPITAL, LONDON.

867.   R.   Glycerini acidi tannici,   f ℥iij.
          Olei olivæ,
          Misturæ acaciæ,   aa. f.℥j.   M.

This injection our author has extensively employed in hospital and private practice. It should be used in the following manner: The bladder having been first emptied, the bottle containing the lotion is to be well shaken, and about two drachms of it briskly poured into a saucer, and quickly drawn into a syringe. The penis is then to be held in the left hand, with the thumb and little finger respectively placed upon the superior and inferior portions of that organ, close to the symphysis pubis, and the fore and middle fingers resting in like manner upon the superior and inferior surfaces of the glans, close to the meatus urinarius. The syringe, with the piston withdrawn, is now to be taken up with the right hand, and the nozzle, as far as its shoulder, carefully passed into the urethra. The thumb and little finger must press the root of the penis to prevent the passage of any fluid beyond that point. When a sense of tension is felt, the syringe may be withdrawn; but the front fingers must previously be so applied as to compress the glans, and thus prevent any escape of the fluid. Next, with the thumb and forefinger of the right hand, the fluid in the urethra is to be set in motion, and so kept for four or five minutes. This will be attended with a

---

* Syphilis and Local Contagious Disorders. Am. Ed., 186?. P. 361, et. seq.

gurgling noise, from the mixture of air and fluid. Thus, when the injection has insinuated itself within the folds and lacuna of the urethra, it is allowed to escape. In this manner, it is asserted, the bladder is protected on the one hand, and on the other there is a certainty of the fluid being applied to the whole of the affected surface.

Glycerinum acidi tannici is official in the British Pharmacopœia. It is made by rubbing together in a mortar one ounce of tannic acid and four ounces of glycerine, then transferring the mixture to a porcelain dish, and applying a gentle heat until complete solution is effected.

M. Luc, a French military surgeon, uses in gonorrhœa, when the discharge is without pain, an injection of a thin paste of finely powdered starch and hot water.

J. JENNEL, PROFESSEUR A L'ÉCOLE DE MÉDECINE DE BORDEAUX, ETC.

Our author prefers to employ for injections an emulsion of copaiba instead of the water distilled with copaiba, recommended by Prof. LANGLEBERT (p. 282). The following is his formula, which may be diluted with water to obtain weaker emulsions, which are all perfectly stable.

OFFICIAL STANDARD EMULSION OF COPAIBA.

868.   R.   Copaibæ,   ℥x.
            Pulveris sodæ carbonatis
                (cryst.),   ℥xx.
            Aquæ destillatæ,   f.℥xxx.   M.
        Mix in a four pint bottle and shake.

This mixture forms a perfectly homogeneous emulsion, which remains so for several hours. The copaiba separates by the next day, but is emulsioned again by the slightest agitation.

ANTI-GONORRHŒAL INJECTION OF COPAIBA.

869. R.   Emulsion. copaibœ (F. 868), f.ʒvj.
           Vini opii,   gtt.xij.
           Aquœ destillatœ,   f.ʒxviij.   M.

An emulsion is thus obtained whose stability is indefinite, and which contains one per cent. of copaiba, and one-half of one per cent. of carbonate of soda.   This formula given as a type may be varied according to indications.   The efficacy of this injection has been proven by a long experience at the military hospital at Bordeaux.*

PROF. EDMUND LANGLEBERT, PARIS.

ABORTIVE INJECTION.

The nitrate of silver, in a slightly caustic solution, is the only article which suits for this sort of injection.

870. R.   Argenti nitratis,   gr.j–ij.
           Aquœ destillatœ,   f.ʒj.   M.

The injection ought to be made with a syringe, with a re-current jet, so as to cauterize only the anterior part of the urethra to the depth of about two inches.   There is danger in injecting more deeply.†

ASTRINGENT INJECTIONS.

A great number of articles are used in these injections.   The principal are the sulphate of zinc, the sulphate of copper, the nitrate of silver, alum, the *pierre divine* (lapis divinus), the iodide and per chloride of iron, tannin and its compounds.   As vehicles, distilled water is generally employed, or rose water ; but it is preferable, except for the nitrate of silver, to make use of *water distilled with copaiba*, thus giving injections which act both by the virtue of the substances they contain, and by the fluid employed to dissolve them.

*Annuaire Pharmaceutique par L. Parisel, Cinquième année, p. 131.

†Aphorismes sur les maladies vénériennes, par Edmund Langlebert, Professeur, etc, Paris, 1868, p. 149.

871. ℞. Zini sulphatis, gr.vijss.
Aquæ copaibæ, f.℥iv. M.

872. ℞. Zinci sulphatis, gr.vijss.
Vini opii, ℳ.xv-xxx
Aquæ copaibæ. f.℥iv. M.

873. ℞. Zinci sulphatis, gr.vijss.
Morphiæ muriatis, gr.¾–jss.
Aquæ copaibæ, f.℥iv. M.

874. ℞. Zinci sulphatis, gr.vijss.
Atropiæ sulphatis, gr.¾–jss.
Aquæ copaibæ, f.℥iv. M

875. ℞. Zinci sulphatis, gr.vijss.
Cupri sulphatis. gr.¾–jss.
Aquæ copaibæ, f.℥iv. M.

876. ℞. Zinci sulphatis, gr vijss.
Lapidis divinii, gr.¾–jss.
Aquæ copaibæ, f.℥iv. M.

(The French codex gives the following formula for the *Lapis divinus*.)

877. ℞. Cupri sulphatis (cryst.)
Potassæ nitratis,
Aluminis, aa. ℥iij.
Camphoræ, ℨj.

Reduce the first three articles to a powder, place in a crucible, and heat so as to undergo the watery fusion ; add the camphor reduced to a powder and permit the mass to congeal on an oiled stone. (When the mass is cold break it up into pieces, and keep in a well stopped bottle.*)

878. ℞. Zinci sulphatis, gr.vijss.
Tincturæ catechu, ℳ xv-xxx.
Aquæ copaibæ, f.℥iv. M.

879. ℞. Zinci sulphatis, gr.vijss.
Tincturæ krameriæ, ℳxv.
Aquæ copaibæ. f.℥iv. M.

880. ℞. Zinci sulphatis, gr.vjss.
Aluminis, gr.iv.
Aquæ copaibæ, f.℥iv. M.

*Pharmacopœia Francaise, p. 650.

881. ℞. Lapidis divini,   gr. iv·
     Vini opii,   m̶. xv.
     Aquæ copaibæ,   f.℥iv.   M.

882. ℞. Zinci sulphatis,
     Plumbi acetatis,   aa· gr. vijss.
     Tincturæ catechu,   m̶xv.
     Aquæ copaibæ,   f.℥iv.   M.

883. ℞. Argenti nitratis,   gr. ¾–jss.
     Aquæ destillatæ,   f.℥iv.   M.

The above injections may be employed in all the stages of acute urethritis. It should be borne in mind that the dose of gr. vijss. of sulphate of zinc given in these formulæ, is only the mean from which the strength is to be increased or lessened according to the indication. The injection should not be of such a strength as to excite pain, otherwise the inflammation will be augmented and continued instead of being relieved.

The following injections are principally recommended in the stage of decline of gonorrhœa :

884. ℞. Tincturæ iodinii,   gtt. xv–xx.
     Aquæ copaibæ,   f.℥iij.   M.

885. ℞. Ferri iodidi,   gr. iij–ivss.
     Aquæ copaibæ,   f.℥iij.   M.

886. ℞. Ferri chloridi,   gr. iij–ivss.
     Aquæ copaibæ,   f.℥iij.   M.

887. ℞. Cupri sulphatis,   gr. iij–ivss.
     Aquæ copaibæ,   f.℥iij.   M.

888. ℞. Acidi tannici,   gr. xv.
     Aquæ copaibæ,   f.℥iij.   M.

889. ℞. Extracti krameriæ,   gr. xv.
     Aquæ copaibæ,   f.℥iij.   M.

890. ℞. Hygragyri chlor. cor.,   gr. ¼–iss.
     Aquæ destillatæ,   f ℥iij.   M.

891. ℞. Atropiæ sulphatis,   gr. iss–iij.
     Aquæ destileatæ,   f.℥iv.   M·

892. ℞. Zinci tannatis,   ʒss–Ɖiv.
    Aquæ copaibæ,   f.℥iij.   M.

893. ℞. Bismuthi subnitratis,   ʒss–Ɖiv.
    Aquæ copaibæ,   f.℥iij.   M.

894. ℞. Pulveris catechu,   ʒss–Ɖiv.
    Aquæ copaiba,   f.℥iij.   M.

895. ℞. Zinci sulphatis,   gr. vijss.
    Zinci oxidi,   ʒss–Ɖiv.
    Aquæ copaibæ,   f ℥iv.   M.

The last four days of the above injections are principally designed to carry into the urethra pulverized matter, which, disposing itself upon the walls of that canal, prevent their contact.

It is necessary to shake the liquid before using it in order to place the powder in suspension.

The following ointments may be used to cover bougies of wax or rubber introduced into the urethra in obstinate cases of gonorrhœa:

896. ℞. Argenti nitratis,   gr. xv–xxx.
    Adipis,   ℥j.   M.

897. ℞. Acidi tannici,   ʒj.
    Adipis,   ℥j.   M.

898. ℞. Hydrargyri chloridi mitis,   ʒss.
    Adipis,   ℥j.   M.

899. ℞. Potassi iodidi,   ʒj.
    Adipis,   ℥j.   M.

900. ℞. Extracti belladonnæ,   Ɖiv.
    Adipis,   ℥j.   M.

This kind of catheterism is not always inoffensive, and should be used only with great care. The same is true of the cauterization of the urethra, by the solid nitrate of silver. Useful sometimes, these means are oftener ineffectual and can cause grave complications.

BALSAMIC CONFECTIONS.

It should be remembered that the copaiba and cubebs, which form the base of these preparations, ought only to be administered at the moment when the acute symptoms of the urethritis commence to subside; that is to say, when the pain and inflammatory swelling have nearly disappeared.

901. ℞. Copaibæ,   f.ʒvss.
    Cubebæ,   ʒxj.
    Catechu,   Əij.
    Tr. olei menthæ piperitæ, gtt.x.   M.
Fiat electuarium.

902. ℞. Copaibæ,   f.ʒj.
    Cubebæ,   ʒij.
    Ferri carbonatis,   Əij.
    Syrupi,   q. s.   M.
Fiat electuarium.

903. ℞. Copaibæ,   f.ʒj.
    Cubebæ,   ʒxj.
    Catechu,   ʒj.
    Magnesiæ,   Əij.
    Tr. olei menthæ piperitæ, gtt.x.   M.
Fiat electuarium.

904. ℞. Copaibæ,   f.ʒij.
    Cubebæ,   ʒxj.
    Magnesiæ,   ʒj.
    Camphoræ,   ʒss.
    Tr. olei menthæ piperitæ, gtt.x.   M.
Fiat electuarium.

These various confections ought to be taken three times a day, in the dose of a drachm or a drachm and a half each time, in some unleavened bread. Or they may be administered under the form of sugar-plums or of capsules.

BALSAMIC MIXTURES, PILLS AND POWDERS.

905. ℞. Copaibæ,
    Alcoholis,
    Syrupi tolutanus,   aa. f.ʒij.
    Aquæ menthæ piperitæ, f.ʒiv.
    Spiritûs ætheris nitrosi, f.ʒij.   M.
From three to six dessertspoonfuls during the day.

906. ℞. Copaibæ, ℥j.
Acaciæ, ℥ijss.
Aquæ aurantii,
Aquæ lactucæ, aa. f.℥jss.
Syrupi, f.℥vj. M.

Three to six dessertspoonfuls per day.

907. ℞. Copaibæ, f.℥jss.
Syrupi tolutanus, f.℥vj.
Pulveris acaciæ, ℥ijss
Tinct. olei menth. piperitæ, gtt. xij. M.

Four to six dessertspoonfuls per day.

908. ℞. Pulveris cubebæ, ℥iij.
Sodæ bicarbonatis, ℈iv. M.

For twenty powders; take from four to six a day in some unleavened bread or in water. This mixture can equally well be ordered in capsules.

909. ℞. Pulveris cubebæ, ℥iij.
Pulveris aluminis, ℥ss–j. M.

To be given in the same manner as the preceding.

When copaiba and cubebs are not borne by the stomach, they may be given, according to the following formulæ, in the form of

ENEMATA.

910. ℞. Copaibæ, f.℥ss.
Ovi vitelli, j.
Extracti opii, gr. ⅓.
Aquæ, f.℥viij. M.

911. ℞. Copaibæ, f.℥ss.
Camphoræ, gr.vijss.
Ovi vitelli, i.
Aquæ, f.℥viij. M.

912. ℞. Pulveris cubebæ, ℥iv.
Vini opii, gtt.x.
Infusi lini, f.℥viij. M.

BALANO-POSTHITIS.

Balano-posthitis is, of all the venereal maladies, the easiest to cure. Astringent lotions or injections suffice, in the greater proportion of cases, to cause its prompt disappearance.

When it is possible to uncover the gland, make three or four dressings a day with a piece of fine linen or lint (inserted between the gland and prepuce), wet with one of the following

ASTRINGENT SOLUTIONS.

913. R. Argenti nitratis,   gr.iij.–ivss.
          Aquæ destillatæ,   f.℥iv.   M.

914. R. Aluminis,   ℈ij.–iv.
          Aquæ rosæ,   f.℥iv.   M.

915. R. Acidi tannici,   gr xv–xxx.
          Vini aromatici,   f.℥xj.
          Aquæ rosæ,   q. s. ad f.℥iv.   M.

(For formula for the vinum aromaticum, see F. 943.)

916. R. Tincturæ iodinii,   ℥xv.–xxx.
          Aquæ destillatæ,   f.℥iv.   M.

Balano-posthitis requires most frequently only local treatment. When, however, the inflammation tends to become phlegmonous, and threatens to terminate in gangrene, it is well to subject the patient to a severe regimen, and to the use of antiphlogistics, diet, repose, general baths, demulcent drinks, saline purgatives, etc. In order to combat gangrene, order

917. R. Camphoræ,   ℥ss.
          Extracti opii,   gr.iij.
          Moschii,   gr.vijss.   M.
For forty pills ; from six to ten a day.

The penis should be wrapped up in compresses, moistened with the following strongly opiated solution :

918. R. Extracti opii,   ℈ij.
          Vini aromatici,   f.℥iij.
          Aquæ rosæ,   f.℥vj.   M.

After the inflammation has subsided, lotions and intra-preputial injections, with the solutions given above, should be resorted to.

In the treatment of vulvitis, our author uses the same remedies, with some slight modifications, as those he employed in balano-posthitis, just given. When the urethra is involved he treats as in men, by astringent injections, copaiba, cubebs, camphor, etc.

VAGINAL INJECTIONS.

919. ℞. Acidi tannici,   ℨijss-xiv.
           Aquæ,   Oij.   M.

920. ℞. Extracti krameriæ,   ℨiv-vj.
           Aquæ,   Oij.   M.

921. ℞. Aluminis,   ℨijss-vj.
           Aquæ,   Oij.   M.

922. ℞. Zinci sulphatis,   ℨijss-vj.
           Aquæ,   Oij.   M.

923. ℞. Ferri sulphatis,   ℈iv-ℨvj.
           Aquæ,   Oij.   M.

924. ℞. Plumbi acetatis,   ℈iv-ℨvj.
           Aquæ,   Oij.   M.

925. ℞. Argenti nitratis,   gr.xv-xxx.
           Aquæ destillatæ,   Oij.   M.

926. ℞. Potassii iodidi,   ℨss.
           Tincturæ iodinii,   f.ℨjss-vj.
           Aquæ,   Oij.   M.

927. ℞. Acidi tannici,   ℈iv.
           Tincturæ iodinii,   f.ℨijss.
           Aquæ,   Oij.   M.

928. ℞. Aluminis,   ℈iv-ℨijss.
           Decocti quercus albæ,   Oij.   M.

929. ℞. Ferri iodidi,   ℈iv-ℨijss.
           Aquæ destillatæ,   Oij.   M.

930. ℞. Ferri chloridi,   ℈iv-ℨijss.
           Aquæ,   Oij.   M.

931. ℞. Lapidis divini,   ℨijss.
           Aquæ,   Oij.   M.

(For formula for lapis divinus see F. 877.)

19

932. ℞. Liquoris sodæ chlor.,   f.℥vj.
           Aquæ,   f.℥xxv.   M.

These various solutions ought to be *deeply* injected into the
vagina. Tampons of charpie can be inserted into the
vagina, saturated with the above liquids or the following
mixture:

933. ℞. Acidi tannici,   ℨj.
           Glycerinæ,   f.℥j.   M.

Erosions and granulations of the uterine neck should be
cauterized with nitrate of silver, acid nitrate of mercury or
the tincture of iodine.

COMPLICATIONS OF GONORRHŒA.

934. ℞. Plumbi acetatis,   ℈iv.
           Aquæ,   Oj.   M.

In edema of the prepuce   Or,

935. ℞. Aluminis,   ℨvjss.
           Aquæ,   Oj.   M.

The œdematous organ is to be enveloped and lightly compressed by a linen bandage saturated with one of the above
solutions.

M. SANDRAS.

936. ℞. Aloes,   gr.jss.
           Extracti lactucarii,   gr.ss.
           Glycyrrhizæ,   q. s.   M.

For one pill once or twice a day as a revulsive in gleet.

This recipe has also been used by M. Reynaud (of Toulon)
in acute gonorrhœa with favorable results.*

MR. SKEY, SURGEON TO ST. BARTHOLOMEW'S HOSPITAL, ETC.

937. ℞. Ferri et quiniæ citratis,   ℥ss.
           Syrupi aurantii,
           Aquæ,   aa. f.℥ij.   M.

A teaspoonful ter die in water.

Our author employs this ferruginous preparation internally, together with the local use of mild injections.

---

*Cullierier's Atlas of Venereal Diseases, p. 106.

THOMAS HAWKES TANNER, M. D., F. L. S., ETC., LONDON.

938. R. Liq. plumbi subacetatis, f.ʒj.
Aquæ, f.ʒiv. M.

This injection, our author thinks, is generally the most useful; it should be employed every eight or twelve hours. If it loses its effects after a few days, sulphate of zinc (gr. ij· to f.ʒj.), or chloride of zinc (gr.j. to f.ʒj.) ought to be substituted.

An experienced surgeon, U. S. A., employs the following :

ELECTUARY.

939. R. Pulveris cubebæ, ʒviij.
Magnesiæ,
Aluminis sulphatis, aa. ʒj.
Copaibæ, q. s. to form mass.

The electuary has stood the test of the most enlarged army experience and has always proved a true antigonorrhœal remedy.

The usual dose during the acute stage of the disorder is one drachm, taken four to six times daily, and continued in this way so long as the urine possesses irritating qualities. As the inflammation subsides, three to four times daily will suffice.

Cubebs generally proves acceptable to the stomach and acts as an astringent along the whole genito-urinary mucous track. The alumina, during the acute stage, serves to change the urine from a highly acid condition to a neutral, bland and unirritating fluid.

The magnesia acts as a gentle aperient, overcoming the natural torpor of the colon and rectum, removing all excrementitious material, which might produce sympathetic disturbance of the urinary structure. It is believed that no pathological view of this disease can be correct, which is irrespective of the condition of the alvine canal.

The small quantity of copaiba is always tolerated by the stomach, and, as a renal depurant, assists in an increased elimination of the renal secretions. As experience is the crucible in which the action of all remedies must be finally tested, this formula is unhesitatingly recommended after long use. No disease proves milder in its first attack than gonorrhœa, and none more obstinate and harrassing if neglected— like a hay rick on fire—a grasp of the hand at first may extinguish it—neglected, destruction is inevitable.

Along with the internal remedies suggested, it is important to notice the use and abuse of injections in this complaint. Our author uses only those of a mild and unirritating kind, viz.: either the acetate or sulphate of zinc (2 grs. to the ounce of water), used six or eight times daily during the acute stage—thrown well into the urethra and allowed to remain long enough to come in contact with the follicles and mucous glands of the affected parts. The attempt to carry the citadel of the enemy by storm—by the use of nitrate of silver—must fail to stifle the inflammatory element, and produce, as it too often does, a train of symptoms endangering the safety of the epididymis and testicles, if not inducing organic stricture. The treatment recommended, with the constant use of the local bath for a few days, will so far complete the cure as to entitle the electuary to a prominent place in the list of anti-gonorrheal remedies.

SYPHILIS.

CHANCRES.

FREEMAN J. BUMSTEAD, M. D., NEW YORK.

940. R. Hydrarg. chlor. mitis,   gr.xxxvj.
           Tincturæ opii,   f.ʒj.
           Cerati simplicis,   ʒʒ.   M.

For application to chancre when an unctious dressing is required. It is much used in French hospitals. Unguents are less desirable than lotions, and should only be employed when the evaporation of a water-dressing cannot be prevented, even with the assistance of oiled silk and glycerine, as may happen from the position of the sore, and during a journey, etc.

In most cases the lotion may consist of simple water or glycerine. When medicated, such ingredients should, as a general rule, be added as will not leave a deposit, or change the aspect of the sore, and thus render its condition obscure. The following may be used:

> 941. R.   Acidi nitrici dil.,   f.ʒj.
>   Aquæ,   f.℥viij.   M.

The strength may be varied with the sensibility of the part. When the sore is situated upon the external integument, the dressing should be covered with oil silk.

Chancres located beneath the prepuce may be dressed with dry lint, which will be sufficiently moistened by the natural secretion of the part. Indurated chancres, are not liable to give rise to successive sores in the neighborhood, and hence astringents and disinfectants are rarely required. When the chancre assumes an excavated form, as is commonly seen in the furrow at the base of the glans, scraped lint is preferable to dry linen, since it is a better absorbent.

The frequency with which local applications are to be changed must be determined by the amount of the secretion. A second dressing should be substituted before the first is soaked with the discharge. The dressing of the most uncomplicated chancres need be renewed only two or three times a day, but phagedenic ulcers require a much greater frequency.

> 942. R.   Ferri et potass. tart.,   ℥ss.
>   Syrupi,
>   Aquæ,   aa. f.℥iij.   M.
>
> From two teaspoonfuls to a tablespoonful three times a day, within an hour after meals, in phagedenic chancres, and a lotion containing the same salt to be applied to the ulcer.

RICORD calls this preparation the "born enemy" of phagedena.

M. A. CULLERIER, SURGEON TO THE HOSPITAL DU MIDI, PARIS.

In soft chancre, cauterization or excision should be employed at the commencement of the ulceration, before it is
well established. The effect of cauterization, even after the
chancre has existed several days, is always to check its progress. If the sore shows no destructive tendency, if there is
nothing to indicate any troublesome complication, if inflammation or œdema is present, or if we are dealing with a pusil-
lanimous patient, we may have recourse to another mode of
treatment. This consists in complete repose of the part, scru-
pulous attention to cleanliness, and the use of narcotic, emollient, or slightly astringent lotions. Dry lint is often used for
the purpose of absorbing the pus, care being taken to insert
the lint between the secreting sufaces, on account of the great
facility with which the pus of a soft chancre is inoculated.
The dressing of soft chancre with salves is a detestable mode
of treatment, and often itself the cause of serious complications.
If the suppuration is copious, antiseptics and detergents should
be employed to combat and neutralize in some degree its virulent property—as, for example, the diluted tincture of iodine,
a solution of chlorine, a decoction of cinchona, and especially
aromatic wine, in form of lotions frequently repeated, or by
moistening a bit of lint with the same, and applying it to the
ulceration.

The following formula is given by BUMSTEAD, as a substitute for the aromatic wine of the French pharmacopœia, when
it cannot be procured :

943.   ℞.   Claret wine,
              Spts. lavandulæ comp., aa.   f.℥jj.
              Tincturæ opii,   f.℥j.
              Acidi tannici,   gr.xv–℥ij.
              Aquæ,   f.℥vij.   M.

The dressing should be renewed several times a day.

CULLERIER is a great advocate for an entirely dry dressing. He often advises applications of calomel or alum powder. These substances soon modify the purulent secretion. They favor, however, particularly alum, the exudation of blood, which sometimes necessitates their suspension.

Soft chancre needs merely local treatment in its ordinary evolution. It is only in cases of inflammatory, gangrenous, or phagedenic complications that general treatment is called for. Specific treatment with mercury CULLERIER is entirely opposed to in soft chancre. He also advocates an exclusively local treatment in hard chancre, waiting until the first appearance of the secondary symptoms on the skin and mucous membranes before administering mercury. A dressing of a slightly stimulating character, as with aromatic wine or dry lint, frequently repeated, in addition to the most careful cleanliness, will suffice in most cases of indurated chancre. If there is any irritation, opiated cerate, an ointment containing calomel, or powdered calomel itself, will have a good effect. The patient should also be placed in a good hygienic condition. Under such circumstances the chancre will follow its regular course toward cicatrization, rather slowly, but quite as rapidly as by any other treatment, whether internal or external.

SILAS DURKEE M. D., ETC., BOSTON.

Abortive Treatment of Chancre.—If, as the result of contagion, or of a suspicious connection, the virile organ has upon

---

*Atlas of Venereal Diseases, page 204.

it a papule, pustule, abrasion, or sore, which *may* be the fore-runner of constitutional syphilis, the best thing the surgeon can do, locally, is to make a caustic application to the spot, if this can be done seasonably, say within ten days from the appearance of the abnormal condition. The design of this operation is two-fold: to destroy morbid structure and to create a healthy recuperative action in the part. Our author employs for this purpose *potassa fusa*, the *acid nitrate of mercury* or *concentrated nitric acid*. He never uses nitrate of silver or Vienna paste.

In cases of abrasion, he generally applies *nitric acid* by means of a small bit of lint secured to a silver probe, or, if the surface be very small, by means of the end of a glass rod. The sore is to be freely covered with the acid, warm water being at hand to wash off any excess immediately. The *acid nitrate of mercury*, when used, is applied in the same manner. The slough will be detached in three or five days, and a healthy granulating surface appear. If a solitary vesicle, pimple or pustule is to be destroyed, he sometimes selects *potassa fusa*, which penetrates deeper than either of the liquids mentioned. The end of the stick is reduced to a point and brought in contact with the apex of the morbid growth, or, what is better, break the dome of the pimple with a probe, and empty it of its contents before applying the potassa. To ascertain precisely the work done by the alkali, remove the *debris* or portion destroyed by means of the point of the probe. As the operation is painless, no haste is required, but caution and exactness are both necessary. It is difficult to preserve the solid stick of potassa in a dry state, therefore, it had better be applied placing it on the end of a pointed glass rod or pen. A drop of vinegar will neutralize any super-abundance of the caustic. The extent of surface destroyed by this corrosive substance is about twice as great as it appears to be at the time of its application; the same is also true in regard to the depth to which it penetrates.

As the risk of increasing the inflammatory tendency is small, a moderate degree of inflammation co-existing with the pustule or sore need not prevent cauterization.

Cold-water dressing, or a soft cracker poultice may follow the use of the caustic for two or three days. The first is to be preferred. The patient should rest and diet. When the eschar has separated, dress with:

944. R. Ferri et potassæ tartratis, ℈ij.
  Aquæ,   f.℥viij.   M.
  To be applied on lint.
Nitric acid (gtt.ij. to f.℥j. aquæ) makes a clean and suitable dressing also.

If the purulent discharge be abundant, order:

945. R. Acidi tannici,   gr. xv.
  Vini aromatici,   f.℥iij.   M.
  (For vinum aromaticum see F. 943).

If the sore becomes painful, lay over it a piece of lint soaked in

946. R. Extracti opii,   ℈ij.
  Aquæ,   f.℥iv.   M.

In occasional instances, after the application of the caustic and the after-dressings mentioned, the sore assumes a spongy or fungoid aspect. Then apply

947. R. Acidi tannici,   ℈j.
  Tincturæ lavandulæ,   f.℥ss
  Vini rubri,   f.℥iv   M.*

CONSTITUTIONAL TREATMENT OF CHANCRE.

Our author is partial to the use of corrosive sublimate internally in the treatment of indurated chancre. He advises its use in pill form :†

---

*Treatise on Gonorrhœa and Syphilis, 5th ed., p. 196.
†Ibid. p. 224.

948. ℞. Hydrargyri chlor. corros.,
    Ammoniæ muriatis,   aa. gr. xvj.
    Aquæ destillatæ,   f.ʒjss.   M.

Make a solution and make up with bread crumbs,
into cxxviij pills.

This formula gives one-eighth of a grain of corrosive sublimate to each pill. One to be taken morning and night,
immediately after meals. In five or six days one may be
taken ter die. If pills cannot be taken, order

949. ℞. Hydrargyri chlor. corrosivi,
    Ammoniæ muriatis,   aa. gr. vj.
    Tinc. cinchonæ compositæ, f℥ij.
    Aquæ,   f.℥iv.   M.

A teaspoonful, morning and evening, for one week; afterward ter die, directly after eating. When this medicine
has been taken for twelve or fifteen days, it is good practice to omit it for four or five days, and then resume it.

PROF. S. D. GROSS, PHILADELPHIA.

950. ℞. Ung. hydrarg. nit.,   ʒj.
    Cerati simplicis,   ʒvj–℥j.   M.

In the treatment of chancre no remedy is so efficacious as
this. The objection made to greasy applications can only be
considered as having any force when there is a want of cleanliness. The dressings should be changed every five or six
hours, and care should be taken that the ointment shall always
be very fresh. When the parts begin to granulate, apply

951. ℞. Cerati zinci carbonatis, ʒj.
    Adipis,   ʒvj.   M.

Or, merely a bit of dry lint carefully interposed between
the contiguous surfaces often promotes cicatrization with remarkable rapidity.

PROF. RICORD, PARIS.

952. ℞. Ferri et pot. tart ,   ʒij–viij.
    Aquæ,   f.℥vj.   M.

This solution is much employed for the treatment of phage-denic chancres.

BERKELEY HILL, M. B., LONDON, F. R. C. S., ETC.

Our author states that in the treatment of soft chancres, the first thing is to remove general causes of irritation, such as too stimulating diet, wine, and especially venery. All severe exercise must be relinquished—in fact, confinement to the house for some days is often time gained by the progress the sore makes with rest. While the wound is healing, the patient should always avoid standing long at a time, to lessen the risk of bubo; the horizontal position, moreover, greatly pro-motes healing of the sore. If erections at night are trouble-some, they may often be prevented by the patient's last meal being a light one, taken two or three hours before bed-time. For persons of ordinary health it is not necessary to do more than this, but if patients are exhausted or in a debilitated condition, ordinary rules for improvements of the health are necessary; quiet, rest, with good diet, and stimulants, must be freely given. The digestion may be invigorated by tonics, such as

953. R. Acidi nitrici diluti, f.ʒj.
Extracti cinchonæ fluidi, f.ʒij. M.
From xxx to xlv drops in water ter die.

Or,

954. R. Tincturæ ferri chloridi,
Spiritûs chloroformi,
Glycerinæ, aa. f.ʒj. M.
A teaspoonful ter die in water.

LOCAL TREATMENT OF THE SORES.

Most sores need only cleanliness to allay irritation and induce them to granulate. The sore should be washed three or four times a day while the discharge is abundant, and covered with pieces of lint dipped in cold water, over which oil silk should be wrapped, if the sore is situated in an outward part, like the dorsum penis or groin. If the patient is a man, he should be directed to support the penis in a suspensory bandage or handkerchief against the abdomen, never to let it hang down, and to be particular that the dress is loose enough not to chafe the parts in walking. If the sore is underneath the foreskin, the lint should be so interposed that the skin does not touch it, both to prevent the sore being chafed and to avoid the formation of fresh ulcers. Care is particularly necessary in women, whose genital organs are difficult to dress. Strips of lint should be laid between the labiæ on each side, and in the folds of mucous membrane around the vagina. A pledget of cotton-wool dipped in some weak astringent may be placed in the entry to the vagina. The œdema of the vulva, which is common with chancres, is best managed by allaying the irritation with frequent washing and by lying down.

If the sore is indolent and shows no tendency to heal, it should be dressed with some weak astringent solution ; diacetate of lead, sulphate of zinc, or nitrate of silver, from one to four or five grains to the ounce of water ; or, a lotion of five or ten grains of tannin to the ounce of water, with a little red wine ; or, black or yellow wash, are all useful in stimulating the sore to granulate, if the first applications do not answer. Solutions of a caustic strength should not be continually applied, as they only increase the irritation and spread the sore. When used at all, they should be strong enough to produce an eschar at once. Creeping, sluggish sores are often induced to granulate freely by dressing them with

955.   R.   Ferri et potassæ tartratis,   gr.v–x.
          Aquæ,   f.ʒj.   M.

This application is very effectual even in the most obstinate
sores, causing them to granulate and cicatrize rapidly, when
many other remedies have failed. When the sore is very indolent, neither spreading nor healing, calomel or red precipitate may be dusted over the surface. Or,

956. R. Hydragyri sulphatis rubri, gr.ss.
Adipis, ʒij. M.
This ointment is to be laid on for one or two hours.

When the ointment is removed, a lotion of sulphate of
zinc (gr. ij. to f.ʒj. aquæ) may be used to dress the sore. If
means of this kind fail to excite cicatrization, it is better to
destroy the surface thoroughly with caustic to procure fresh
granulations when the eschar separates.

Caustic should be used in the following cases: It may be
employed in the very first appearance of a sore, to shorten its
duration, and to prevent the danger of inflammation, sloughing, multiplication by consecutive inoculation, or bubo. At
this time the sore is also very small and the pain accompanying its extirpation not very great. But when the patient has
had the ulcer a week or ten days before he comes under treatment, the chancre has generally assumed the character it
means to preserve; if it appears little prone to spread and in-
flame, it may be managed by astringent lotions, without resorting to caustics. If the sore, on the other hand, be spreading with sharply cut edges, or if it has lasted a long time, and
resists other treatment; or, again, if its presence prey upon
the patient's spirits, cauterization is the best remedy to prevent further mischief. In the rapidly sloughing chancre with
inflammation, complete cauterization with hot irons is the
most effectual remedy; but it must be followed by soothing
applications, to allay the pain and inflammation when the
sloughing surface is destroyed.

As chancres may excite bubo at any period of their existence, destruction of their surface with caustic may prevent
this consequence whenever it is employed. Still, this advantage is not sufficient in practice to require the invariable use
of caustics, as the chance of a particular sore not being accompanied by bubo is two to one, even when left to run its
course. Besides this, it is often exceedingly difficult to destroy
several sores thoroughly by one application of caustic; hence
the patient, after having undergone all the suffering and inconvenience of cauterization, may be disappointed on finding
in a few days his sore assume its original character.

Several preparations are used to destroy the ulcerating tissue. Whichever caustic is selected it should always be
thoroughly applied, and it is better to cauterize a little more
deeply than is absolutely necessary, that complete destruction
of the sore may be insured. Among the most effectual caustics is one RICORD prefers. He makes a paste of powdered charcoal and strong oil of vitriol, which he lays on and
rubs into the chancre. In a few minutes the surface is destroyed, and forms an eschar or crust which falls off in a
week, leaving the sore a simple granulating surface. It is a
very effective remedy, being not liable to overflow the sides
of the ulcer and attack the healthy skin, as is the case with
liquid caustics. But it is not always at hand, hence less convenient than another—the *strongest nitric acid*. The best way
to use this is to daub it with a glass brush over the floor and
edges of the ulcer, and allow it to soak well into the surface
of the sore for a few minutes, before the excess of acid is neutralized with a little carbonate of soda dissolved in water.
The skin surrounding the ulcer should be protected by grease,
but the edges may be left clear for the action of the caustic.
The chloride of zinc and caustic potash are slower in action,

and must be left longer in contact with the sore, or they will
not penetrate deeply enough to destroy it altogether. The
actual cautery, by hot iron or galvanic wire, is at times very
useful when a large amount of tissue has to be destroyed;
otherwise it is not preferable to chemical caustics, while it
alarms the patient much more than the latter. When the
caustic has done its work and the excess washed away with
cold water, the sore should be wrapped in wet lint, and the
pain, which often lasts several hours, can be assuaged by the
constant application of ice-cold water. The eschar usually
separates in four or five days, and leaves a clean granulating
surface. These applications are all very painful, and the for-
titude of the patient seldom affords the surgeon full leisure
for the complete destruction of the sore, and unless this is attained, the suffering will be in vain, whence it is best to render him insensible by chloroform or ether spray. The latter
remedy is exceedingly painful if the part is at all inflamed;
in such cases it is best to use chloroform, which has the additional advantage of rendering the patient unaware of what is
going on around him, and prevents the disagreeable consciousness of the nature of the operation.

Caustics must be withheld in inflamed chancres, except
when the destruction of tissue is very rapid, and thick layers
are recrossing one after another. If, however, the chancre is
simply inflamed, that is, painful, hot, secreting much pus, and
the skin round the sore red and tense, but the ulcerating action does not threaten a great loss of tissue, it is better to
allay the inflammation by rest, moderate diet, and cold lotions.

When the destruction of tissue is so rapid as to cause
sloughing phagedena, the sloughing must be arrested by destroying the surface of the sore beyond the limits of the mor-

bid action, which causes the mortification. The patient should be put under chloroform, and the surface of the sore cleared of the loose sloughs by snipping them away with scissors and wiping the surface dry. Then the red hot iron should be passed evenly over the sore, and along its edges, till all the inflamed and ulcerating tissue is charred. This should be done deliberately and carefully, or the sloughing will begin again in a few hours. The pain of the cauterization may be allayed by wrapping the part in rags dipped in ice-cold water for the first few hours after the operation. When the aching has subsided, a warm linseed poultice may be applied to hasten the separation of the eschar and cleaning of the surface.*

PROF. EDMUND LANGLEBERT, PARIS.

The caustics most useful in destroying the virus of chancre are nitric acid, and the paste of Canquoin :

957.   R.   Zinci chloridi,   ℥ij.
           Farinæ,   ℥iv.
           Alcoholis,   q. s.   M.

This paste ought to be applied in the form of a button having the same dimensions as the chancre, to be held in place by means of a bandage, for a half hour or an hour, depending upon the extent and depth of the ulcer.

The following carbo-sulphuric paste is to be applied in the same manner, with this difference, that instead of removing it at the end of a certain time it is to be permitted to dry on the chancre upon which it remains, until the fall of the eschar :

958.   R.   Acidi sulphurici,   ℥ijss.
           Pulveris carbonis ligni,   q.s.   M.

Soft chancres which it has not been profitable to destroy

---

*Syphilis and Local Contagious Disorders. Am. Ed., 1869, p. 388.

by caustics, ought to be dressed several times a day with charpie saturated, according to the indications, with one of the following

ASTRINGENT LOTIONS.

959. R. Aluminiis,   ℈ij–iv.
          Aquæ rosæ,   f.℥ij.

960. R. Vini opii,   ℳxv.–xxx.
          Vini aromatici,   f.℥iij.   M.

(For formula for vinum aromaticum, see F. 943.)

961. R. Extracti opii,   gr.xv.–xxx.
          Decocti cinchonæ,   f.℥iij.   M.

962. R. Acidi tannici,   gr.xv.–xxx.
          Aquæ rosæ,   f.℥iij.   M.

963. R. Argenti nitratis,   gr.xv.–xlv.
          Aquæ destillatæ,   f.℥iij.   M.

964. R. Tincturæ iodinii,   f.℥jss.–ijss.
          Aquæ destillatæ,   f.℥iij.   M.

965. R. Ferri et potassæ tartratis,   ℈iv.–℥v.
          Aquæ destillatæ,   f.℥iij.   M.

966. R. Zinci chloridi,   gr.jss.–iij.
          Aquæ destillatæ,   f.℥iij.   M.

The lotions of potassio-tartrate of iron and of the chloride of zinc are particularly indicated in order to combat *phagedena*. For the same purpose, the following may also be prescribed:

967. R. Pulveris carbonis ligni,
          Pulveris cinchonæ,   aa. ℥ijss.   M

968. R. Creasoti,   gtt.xv.–xlv.
          Aquæ destillatæ,   f.℥x.   M.

969. R. Potassii iodidi,   gr.xv.
          Tincturæ, iodinii,   f.℥iss.–v.
          Aquæ destillatæ,   f.℥iij.   M.

This last recipe is the one which has given the best results in the hands of our author.

20

OINTMENTS.

Pomades and fatty substances generally are not suitable for the treatment of simple chancre. Their employment ought to be exclusively reserved for those cases in which acute *bubo* first appearing it is wished to promote resolution:

970.	R.	Extracti belladonnæ,	gr.xlv–ℨiv.	
		Unguenti hydrargyri,	ʒvj.	M.
971.	R.	Cerati opii,	ʒijss.	
		Unguenti hydrargyri,	ʒss.	M.
972.	R.	Potassii iodidi,	ʒss.	
		Adipis,	ʒj.	M.
973.	R.	Potassii iodidi,	gr.xv.	
		Plumbi iodidi,	℈ijss.	
		Adipis,	ʒiv.	M.
974.	R.	Extracti opii,	gr.xv.	
		Plumbi iodidi,	ʒss.	
		Adipis,	ʒj.	M.

Simple buboes (those from soft chancres) are the only ones in which resolution may be hoped for. When the buboes are chancrous or virulent, they suppurate inevitably. Once they are open, their treatment is the same as that of the chancres from which they come.

The soft chancre ordinarily requires only a local treatment. If, however, the patient be feeble and anæmic it will be well to place him upon chalybeates and bitters.

975.	R.	Ferri iodidi,	℈iv.	
		Ferri redacti,	gr.xv.	
		Extracti gentianæ,	ʒijss.	M.

For xl pills. Two to four a day.

| 976. | R. | Ferri iodidi, | ℈iv. | |
| | | Syrupi sarsaparillæ, | Oj. | M. |

From two to four dessert-spoonfuls a day.

THE HARD CHANCRE AND SPECIFIC ADENITIS.

The astringent lotions just prescribed for the soft chancre are equally suitable for the treatment of the infecting chancre,

when this inflames, suppurates and tends to become phagedenic. In ordinary cases, it suffices to dress the chancre with
some charpie covered with a light coating of a mercurial
ointment.

POMADES AND OINTMENTS.

977.	℞.	Hydrargyri chloridi mitis,	gr.xv.
		Adipis,	℥ss.	M.
978.	℞.	Hydrargyri chloridi mitis,	gr.xv.
		Glycerinæ amyli,	℥ss.	M.
979.	℞.	Hydrargyri iodidi viridis,	gr.xv.
		Adipis,	ℨv.	M.
980.	℞.	Cerati opii,	ℨijss
		Unguenti hydrargyri,	℥ss.	M.
981.	℞.	Hydrargyri chloridi mitis,	ℨss.
		Pulveris opii,	gr.xv.
		Adipis,	℥j.	M.
982.	℞.	Hydrargyri ammoniati,	gr.xv.
		Unguenti aquæ rosæ,	ℨvj.
		Olei rosæ,	gtt.iv.	M.
983.	℞.	Hydrargyri ammoniati,	gr.xv.
		Morphiæ muriatis,	gr.iv.
		Cerati opii,	℥vj.	M.

In the greatest number of cases the sympathetic adenitis of
the infecting chancre does not require any local treatment.
If, however, the ganglionic tumor becomes too large, it will
be well to apply an alterative ointment.*

THOMAS HAWKES TANNER, M. D., F. L. S., LONDON.

This author advocates the use of mercury so soon as induration appears, to be continued until all hardness at the seat
of inoculation, or of the inguinal glands, has entirely gone.
Local applications are of comparatively little use. He directs that from an eighth to a fourth of an ounce of mercurial
ointment should be rubbed into the inside of the thighs every

---

*Aphorismes sur les Maladies Vénériennes. Paris. 1868, p. 171,

night at bed time, until the gums are slightly touched. Or a mercurial vapor-bath should be taken at first every night, and then two or three nights a week.

CONSTITUTIONAL SYPHILIS.

WILLIAM AITKEN, M. D., EDIN.

984. R. Hydrarg. chloridi, corros., gr.j.
Potassii iodidi, gr.xxx.
Liq. potassæ arsenitis, ℩xxxvj.
Alcoholis, f.ʒj.
Ext. sarsaparillæ fluidi, f.ʒiij.
Aquæ cinnamomi, ad f.ʒxij. M.

Two tablespoonfuls three times a day, after meals, in the treatment of some of the more intractable forms of syphilitic squamæ.

JOHN K. BARTON, M. D., (DUB.) F. R. C. S. I., ETC.

Our author recommends mercury as generally necessary in the first and second stages of the disease, though, with Ricord, he believes its action is limited to causing the disappearance of the symptoms present when it is administered to, and that it cannot be considered capable of neutralizing the poison. He lays great stress upon its gradual introduction into the system, and, in common with Colles, Brodie, and Sigmund, prefers that this should be effected by inunction.

The patient's diet and daily habits should in the first place be regulated; the former should consist of meat once daily, without any stimulants beyond beer or porter, sometimes better without any at all. He should keep regular and early hours, going to his bed not later than ten o'clock, and not rising before eight in the morning; during the day he may be engaged in business, if it be not of a laborious or exciting description.

985. R. Unguenti hydrargyri, ℥j.

Of this half a drachm should be rubbed in each morning after breakfast, for twenty minutes or half an hour. The morning is the best time, because then the patient is most vigorous; and beside, if rubbed at night, the heat and perspiration produced by lying in bed will cause a considerable loss of the ointment, and the patient breathes an atmosphere loaded with mercury. Unless the full time mentioned be given to the rubbing, half the ointment will be inefficient. It is usually necessary to impress the importance of this upon the patient, who, however, in a very short time lends a willing aid to the surgeon, finding his symptoms disappearing gradually, and his general health and strength improving rather than decreasing.

The inside of the thigh and popliteal space is the region where the inunction can be practiced. The patient should be told to rub in on each thigh upon alternate mornings, carefully washing off the old ointment with warm soap and water before commencing the new inunction; this prevents the skin becoming irritated, and mercurial eczema appearing; if, however, a few scattered pustules do appear, the rubbing should be transferred to the axillæ for a time. He is in the habit of directing the patient to take a hot-air or Turkish bath once or twice a week during treatment, and finds it not only preserves the skin from irritation by thoroughly cleansing it, but also facilitates the action of the mercury; patients, including those in hospital, always express a sense of comfort and relief from the use of the bath.*

Many cases, particularly those belonging to the first division of the tertiary stage, are most benefited by a combination of mercury and iodide of potassium. For this purpose

---

*The Pathology and Treatment of Syphilis, Dublin, 1868, p. 226.

add to the recipe, gr.1-16—1-12 of the corrosive chloride, or the bin-iodide of mercury to each dose.

When our author employs mercury internally in secondary syphilis, he considers the following a good combination:

> 986. R. Pilulæ hydrargyri,   ℥j.
>   Extracti opii,   gr.v.   M.
>   For xx pills. One of these daily will be as good internal
>   treatment as is possible.

Iron or quinine may at times be advantageously combined with some of the preparations of mercury, particularly when marked symptoms of anæmia show themselves at the commencement of the secondary period, which is very frequently the case in women.

> 987. R. Pilulæ hydrargyri,   gr.xx.
>   Ferri sulphatis exsiccatæ, gr.x.
>   Extracti opii,   gr.v.   M.
>   For xx pills.

> 988. R. Hydrargyri cum cretâ,
>   Quiniæ sulphatis,   aa. ℈j.
>   Extracti opii,   gr.iij.   M.
>   For x pills.

The special treatment for *secondary ulceration of the throat* is

> 989. R. Argenti nitratis,   gr.xxx–xl.
>   Aquæ destillatæ,   f.℥j.   M.

To be freely applied over the velum and back of the pharynx every day, or every other day, while any ulceration or redness continues. The same solution may be used with the spray producer.

If toward the close of the secondary period sore throat reappears, as it often does, it then does not yield so rapidly, and it will be necessary to prescribe the following mixture, which will quickly cause it to heal:

990. ℞. Potassii iodidi, ℈ij.
Potassæ chloratis, ℈iv.
Aquæ, f.℥viij. M.
Two tablespoonfuls ter die.

In the tertiary stage our author employs iodide of potassium
in doses of from eight to ten grains ter die. A salt of am
monia added to the solution seems to increase the activity of
the iodide, thus:

991. ℞. Potassii iodidi, ℥iv.
Ammoniæ muriatis, ℥ij.
Tr. cinchonæ compositæ, f.℥iv. M.
A teaspoonful in a wineglassful of water, ter die.*

W. BŒCK, M. D., PROFESSOR IN THE UNIVERSITY OF CHRIS-
TIANIA, NORWAY.

Our author has practiced the treatment of constitutional
syphilis by

SYPHILIZATION

During seventeen years. He began an unbeliever, but was
forced by experience to abandon his prior convictions.

He considers it unadvisable to commence any inoculation
until secondary symptoms have appeared. Then he commences as soon as possible. He takes his matter, when possible,
from the indurated chancre, which in most cases must be first
irritated in order to yield purulent matter, which alone is in-
oculable. He does not object to the use of matter from the
soft chancre, experience having taught him that the result is
the same.

He takes the matter on the lancet and inoculates precisely
in the same manner as in ordinary vaccination, using, however, great care—since the inoculation of syphilitic virus does
not take with the same readiness as does that of vaccine

*The Pathology and Treatment of Syphilis. Dublin, 1868, pp. 232, 236, 242.

virus. He commences on the sides of the chest, making three inoculations on each side. One would be enough, were it certain that the virus would take effect, but it often occurs that some inoculations have no results. At the end of three days there are generally pustules developed after the first inoculation. He takes matter from these and inoculates in other places on each side, taking care at each succeeding inoculation to inoculate at a sufficient distance from the former pustule, to prevent anything like a union or confluence of the sores. He continues now in the same manner his inoculations on every third day, taking always his matter from the last pustules formed, until the matter has lost all effect. He then takes matter from some other patient, generally from a patient under treatment by syphilization. He continues with this new matter in the same manner as he did with the first. The second matter does not produce sores as large as the first, and cannot be inoculated through so many generations. When this matter also loses its effect, a third selection is tried. It is possible that this will prove effective, but not to any great extent. It is best at this period to commence over both arms. For this purpose new matter can be selected, or the matter can be taken from the most recent pustules of the chest, because the matter which produces no effect on the chest may produce pustules over the arms. He continues now on both arms in the same manner and pursues the same course as he did over the chest; continuing as long as the matter shows any effect. When the effect on the arms appears to be lessening, he commences at the thighs and continues there in the same manner as on the arms, until there is no appearance whatever on any of the inoculated surface of the existence of any matter.

The reason he commences in the sides is, that he never

produces any phagedenic sores in this region, which might occur were he to commence on the arms or thighs. During the course of these inoculations the syphilitic symptoms gradually subside, and, as a rule, have entirely disappeared when the virus fails to produce any further effect. When they have not quite disappeared at this time, they do so shortly afterward without any further treatment.

In every case he carries out syphilization without any reference to internal remedies, and when the treatment is finished the general health is invariably good. The anæmic condition disappears simultaneously with the syphilitic symptoms.

Sometimes the first inoculations do not take effect. This will be ascertained in twenty-four hours. It is principally seen in those who have strong eruptive forms of the disease, and especially seen in children suffering from hereditary syphilis. Then it is necessary to continue inoculating daily, until the virus has taken effect. So, also, as the case advances toward a cure, it is often found necessary to inoculate daily. If the inoculations are not repeated as directed, there is great danger of relapse. On the appearance of any acute disease, the inoculations are stopped and resumed upon convalescence. Inoculations never produce any result during the puerperal state.

Should iritis be observed, he merely applies atropia, but no other treatment, contenting himself with keeping the patient in a clear, light room, since from experience he believes the dark room only increases the severity of the attack.

Syphilization is the method diametrically opposed to mercurialization. By this method our author does not pretend, for a moment, to arrest the disease, or interfere with nature in her own movements (as he contends is done by the mercurialist); on the contrary his efforts are directed toward assisting her in passing through a course, which too often is but imperfectly accomplished, through her forces. Every one who follows syphilization from three to four months will have an opportunity to witness the phenomena attending the elimination of the syphilitic virus from the system. This series of phenomena is finished by the immunity of the patient from anything like syphilitic infection.

Our author asks his *confrères* who may wish to try this method of treatment to follow strictly his course of operating, and not do as many have unfortunately done, cease when the work is only half accomplished. The patient in this way is not only not benefited but placed in a more serious condition than before. Many have attempted to modify the measure according to their own peculiar notions, and, having failed to accomplish the best results, have condemned unjustly the method. Many have feared to undertake it, particularly in private practice. But no treatment is more easily pursued, and none more satisfactory. The patient will tell from day to day of his improvement, and reward the practitioner by the expressions at least of his gratitude.*

M. BOUILHON, PARIS.

Our author suggests a soluble salt of mercury for hypodermic injection—*i. e.*, the *double iodide of mercury and sodium.*

This salt is manageable and safe; while active, it is not irritating to the tissues. He directs the follwing for

HYPODERMIC MEDICATION.

992. ℞. Hydrarg. et sodii iodidi, gr.iij.
Aquæ, f.ʒiijss. M.

Of this solution, 10 gtt. can be injected every other day

*The *American Journal of Syphilography and Dermatology*. Jan., 1870, p. 7, et seq.

After a week or two the amount of the injection may be increased, ten drops at a time.*

Cullerier (successor of Ricord at the Hôpital du Midi, Paris) says, that in order to be effective, iodide of potassium must be given in a sufficient quantity, as for example, gr. xv. to ʒj. in the course of the day. It is almost useless in the early seconday stage in which it can never supersede mercury.

SIR BENJAMIN BRODIE.

Our author prefers inunction for the treatment of syphilis in the infant:

993. ℞. Unguenti hydrargyri, ʒj.
　　　 Adipis, ʒj. 　　　M.

Spread over a flannel roller and bind it around the child once a day. "The child kicks about, and the skin being thin, the mercury is absorbed. It neither gripes nor purges, nor does it make the gums sore; but it cures the disease."

PROF. WILLIAM A. HAMMOND, M. D., ETC., NEW YORK.

994. ℞. Potassii iodidi, ʒj.
　　　 Hydrarg. chloridi corrosivi, gr.vj.
　　　 Aquae, f.ʒxij. 　　　M.

Of this mixture, a teaspoonful may be taken three times a day till the system is well under its influence. Our author prefers this to any other form of conjoining mercury with iodine. Gradually increase the dose, so that after twenty or thirty days a quarter of a grain of corrosive sublimate is taken three times per day. He has never seen salivation induced by the plan, although he has kept it up continuously for six and eight months at a time.

J. M. DA COSTA, M. D., PHILADELPHIA.

995. ℞. Argenti nitratis, gr. |-j.
For one pill, ter die.

---

*Bull. de Therapeutique, 15 avril, 1869.

In the treatment of syphilitic dysentery, commence with a quarter of a grain and gradually increase to one grain of the silver nitrate. At night order

996. ℞. Extracti opii, gr.j.
For a suppository, to be introduced at bed-time.

In visceral syphilis it is better, at least at first, to treat the disease which is present, rather than the cachexia.

997. ℞. Hydrargyri chlor. corrosivi, gr. j.
Potassii iodidi, ℥ss.
Syrupi ferrri pyrophosph.,
Aquæ, aa. f ℥iss. M.
A teaspoonful ter die in syphilitic rupia.

The syrup of the pyrophosphate of iron is an excellent vehicle for the other articles as well as useful itself. *Sulphur baths* form, in cases of syphilitic chronic pustular affections, a very valuable adjunct to internal treatment.

SILAS DURKEE, M. D., BOSTON.

The sulphureo-gelatinous bath is valuable in secondary papular eruptions:

998. ℞. Potassii sulphureti, ℥ij.
Aquæ, C.xxx M.

Add to this solution:

Ichthyocollæ, ℔.ij.
Aquæ bullientis solutæ, ℔.x. M.
999. ℞. Potassii iodidi, ℥ij.
Extracti gentianæ, q.s. M.
For xxx. pills. One ter die.

For the purpose of rendering the iodide of potassium more agreeable and efficient, our author combines it with carbonate of ammonia. The impression upon the stomach and upon the general sensation of the patient is very pleasant. He uses the following:

1000. R. Ammoniæ carbonatis, ℨiss.
Potassii iodidi, ℨiij.
Syr. sarsaparillæ compositi,
Aquæ, aa. f.℥ijss. M.

A teaspoonful, three or four times a day, in a gill of cold
water.

1001. R. Sodii iodidi, ℨj.
Syr. sarsaparillæ compositi,
Aquæ, aa f.℥iiij. M.

A teaspoonful ter die in water.

1002. R. Liquoris arsenici et
hydrargyri iodidi, f.℥j.
Syrupi aurantii, f.℥vj. M.

A teaspoonful, in water, ter die, on a full stomach, in obstinate squamous syphilitic eruptions.*

PROF. S. D. GROSS.

1003. R. Hydrarg. chloridi corros., gr.j.
Potassi iodidi, ℨij.
Syr. sarsaparillæ compos., f.℥iij. M.

Desertspoonful ter die, shortly after meals, in tertiary syphilis.

Prof. Gross almost invariably combines the bichloride of
mercury with iodide of potassium in the treatment of tertiary
syphilis, particularly when the affection is of long standing.
An infirm broken state of the system is no bar to the use of
mercury in this mode of combination; on the contrary, it often
affords the medicine an opportunity for its best display. To
counteract any disagreeable effects of the above recipe, such
as gastric irritation, diarrhœa, etc. (which, however, rarely
ensue), an anodyne, as a small quantity of morphia, or from
five to ten drops of the acetated tincture of opium may be
combined with each dose.

In regard to the dose of iodide of potassium in the treatment of tertiary syphilis, Prof. Gross states that long experi-

---

* Treatise on Gonorrhœa and Syphilis. 5th Ed., pp. 339 and 341.

ence has taught him that while less than ten grains ter die will rarely do much good, there are few cases in which more than this quantity is really ever needed.

With reference to the employment of iodide of sodium and iodide of ammonium as substitutes for iodide of potassium, Prof. Gross sometimes recommends their use in five grain doses. CULLERIER says that the iodide of ammonium gives no better results than the iodide of potassium, and he has abandoned its use. It has been asserted, however, on good authority, that the iodides of sodium and ammonium will sometimes succeed in doses in which the iodide of potassium has failed. (TANNER and others.) They are more nauseous than the iodide of potassium.

Bromide of potassium has been employed in tertiary syphilis recently. Cullerier says no reliance can be placed on this remedy; BERKELEY HILL asserts that in small doses in conjunction with the iodide it increases the energy of the latter very materially. It should be borne in mind in administering the bromide of potassium that it is decomposed by a syrup.

To overcome the disagreeable taste of the iodide of potassium, so often complained of by patients, PAGET says that a mixture of whisky and the compound syrup of sarsaparilla makes the best vehicle.

BERKELEY HILL, M. B., F. R. C. S., LONDON,

Has found that in the administration of the iodide of potassium it is best to begin with two grains dissolved in an ounce and a half or two ounces of liquid, three or four times daily, before breakfast and between meals, and to increase the dose by a grain or two every three days. If the patient feels no benefit from a moderate amount, as is often the case when the disease is of very long standing, larger doses of eight, ten, or twenty grains should be tried, or even much larger doses. Forty grains *ter die* will sometimes quell an obstinate syphilide which has resisted smaller quantities. Still larger quantities than this have been given without ill effect. Usually, however, the risk of iodism may be avoided by combining ammonia or bromide of potassium with the iodide. The aromatic spirits of ammonia or the carbonate of ammonia is an excellent adjunct. Professor Gross also speaks of the advantage of combining carbonate or muriate of ammonia with iodide of potassium.

HYPODERMIC INJECTION.

1004, R. Hydrarg. chlor. corrosivi, gr.¼.
       Aquæ destillatæ   ℥xxx.   M.
For one injection.

The patient is thus very rapidly brought under the influence of the drug by much less mercury than is used in any other way. The amount taken into the system can also be exactly measured. Introduced in this way in divided doses of about grains 1-5, it produces mercurialization when about one grain has been injected. This condition is kept to the requisite intensity by the daily injection of gr. ¼. This method has the disadvantage of requiring the attendance of the surgeon, and is disliked for the slight pain it causes; hence, it is only to be recommended where circumstances render it doubtful whether the mercury be taken by the patient, or where, as in severe iritis, it is necessary to put the patient under the influence of mercury as quickly as possible. The subcutaneous injection of mercury is also resorted to by SCARENZIO and HEBRA. The latter injects about gr. 1-40 of corrosive sublimate at a time. Dr. LEWIN, of Berlin, adds morphia to the corrosive sublimate for hypodermic injection in syphilis.

A favorite mixture of our author,* in the late form of the disease, is the freshly formed red oxide of mercury, which he makes according to the following formula :

100.5. R. Hydrarg. chlor. corrosivi,   gr.iij.
            Potassii iodidi,   ℈v.
            Ammoniæ carbonatis,   ℨj.
            Tinct. cinchonæ compositæ,
            Aquæ,   aa. f.ℨiv.   M.
A dessertspoonful ter die, half an hour before meals.

PROF. EDWARD LANGLEBERT, ETC., PARIS.

1006. R. Ammonii iodidi,   gr. ij.
            Acaciæ,   q.s.   M.
For xl pills; take from two to eight each day.

1007. R. Potassii iodidi,   gr. v.
            Extracti gentianæ,   gr. v.
            Pulveris althæ,   q.s.   M.
For xl pills; take from five to twenty each day.

DR. LEWIN, OF BERLIN.

Our author considers the non-mercurial treatment of syphilis as little better than a crotchet. He. has employed the hypodermic injection of corrosive sublimate for more than two years at the Berlin Charité, in over 500 cases.

Owing to the corrosive nature of the fluid, he insists upon the syringe being constantly washed out and its point frequently sharpened. In private practice he keeps a marked canula for each patient. He prefers the back, lateral thoracic region, or buttock, as the place of puncture, because less irritative inflamation ensues; but in iritis the temporal region is preferable. In the great bulk of the cases a solution of 4 grains to the ounce was employed, which, supposing the syringe to hold 15 minims, would give ¼ grain each time. In the very sensitive, from ⅛ to 1-10 grain of morphia may be

*Syphilis and Local Contagious Disorders. Am. Ed. 1860. p. 201. et seq.

added with glycerine. The injections are best performed in the forenoon and afternoon, and, if a very rapid cure is sought, again in the evening. The patient need not be confined to his bed, or in warm weather even to the house, care being taken that he is not exposed to chills. Even when this precaution has been neglected, ill results have seldom followed. The diet need not be much restricted, beyond being somewhat diminished in quantity; but alcoholic drinks should only be taken exceptionally. Great care should be taken in keeping the mouth clean, but moderate smoking may be allowed. The pain caused by the injection is sometimes considerable, especially if it be not performed adroitly, or the patient is very sensitive. In general, he soon becomes accustomed to it. The subsequent irritation, which usually soon subsides, sometimes goes on to inflammation, induration, or suppuration, especially if the injection be too strong or too freely used, some patients being far more susceptible than others. Dr. Lewin, in cases of slight venous hemorrhage, that have occurred among his many hundred injections, has never met with an instance of ill consequences supposed to be due to the introduction of the injected substance into the circulation. He found in his 144 male cases that the average quantity of 2 5-6 grains of sublimate were required to effect a cure, while in those of the cases which had previously undergone no other treatment 3 grains were required. In 356 women 2¼ grains sufficed, i. e. ¾ less than in men.

Summing up his opinions, Dr. Lewin states that preference should be given to this mode, because (1) of the rapidity with which the symptoms disappear; this holding an exact proportion to the quantity of sublimate daily injected. Thus two or three injections per diem of ½ to ¾ grain cured numerous cases of iritis in from five to seven days. In these cases

21

of very rapid cure the patient must keep in doors, and avoid all bodily or mental excitement. (2) The results also are certain and precise. In and out the hospital the author, during two years and a half, has treated 900 cases, exhibiting every variety of symptom and group of symptoms; and in almost all these, even in desperate cases, many of which had been fruitlessly treated by other modes, he has met with the most gratifying results. Syphilitic disease of the bones has offered the greatest resistance, for, although the nocturnal pains have been relieved and the subperiosteal deposits removed, yet the bones themselves did not recover their normal volume. (3) The relapses are small in number and slight in character. The statistical comparison of the results obtained by this and by other means shows that while the relapses after the latter amounted to 81 per cent., those following the injection method were only 31 per cent. (4) Finally, the great convenience of the method, to both physician and patient.[*]

M. LIEGEOIS.

Our author employs the following formula for the hypodermic injection of corrosive sublimate in secondary syphilis:

1008. R. Hydrarg. chlor. corrosivi, gr.iij, 1-10.
Morphiæ muriatis, gr.iss.
Aquæ destillatæ, f.ʒxxiijss. M.
Dose, ⅓. xvss. (=about gr. 1-32 of the sublimate). Ordinarily, no inflammation follows this injection.[†]

DR. FELIX VON NEIMEYER, PROF. IN UNIVERSITY OF TUBINGEN.

1009. R. Hyd. chloridi corrosivi, gr.v.
Pulveris ex. glycyrrhizæ, q. s. M.
For twenty pills. One ter die.

---

[*] *British and Foreign Medico-Chirurgical Review.* London, Oct , 1868. p. 553.
[†] *Arch. Gén. de Méd.* Sept., 1869.

SURGEON W. S. W. RUSCHENBERGER, U. S. N.

1010. R. Hydrargyri iodidi rubri, gr.j.
Iodinii, gr.ij.
Potassii iodidi, ꝫj.
Syrupi sarsaparillæ com., f.ʒxv.
Aquæ, f.ʒj. M.
Tablespoonful four times a day.

PROF. V. SIGMUND, OF VIENNA.

This celebrated specialist has contributed an account of the results of his trials of the subcutaneous injection of corrosive sublimate in the treatment of syphilis in his hospital.*

These have been 113 in number, comprising all the forms and complications of disease. Most of the patients have been females, several of these being pregnant or puerperal women. None of them were younger than 18, and only three above 40, and for the most part they belonged to the working classes. In the majority nutrition had not become impaired through syphilis. In those in whom it was defective this was attributable to tuberculosis, intermittent fever, cachexia, and inveterate syphilis, as also to loss of blood on delivery. Some of the patients had already been under treatment by means of other forms of mercury.

The injection employed was that recommended by Prof. Lewin, of the Berlin Charité, viz:

1011. R Hydrargyri chlo. corrosivi, gr.iv.
Aquæ, f.ʒj. M.
Dose, ♏xv=gr.⅓.

In order to prove successful, the injection must be performed with the greatest care and delicacy, good syringes with very fine and sharp canulæ being chosen. The best places for injecting have been found to be the outer side of the thorax, the abdomen, the upper part of the haunch, and the outer

---

* *Medical Times and Gazette*, London, Oct. 23, 1869, p. 496.

side of the upper arm, while the lower half of the haunch, the lower extremities in general, the back, and the inner side of the arm, are to be carefully avoided. Patients treated by other practitioners have applied to Professor V. Sigmund on account of extensive and tedious infiltrations surrounding the points of injection, and sometimes obstinate ulcerations, and in these cases the injections have usually been made on the back, and in the most troublesome cases on the inner surface of the thigh. In his own clinic he has met with very few cases in which any considerable inflammation was produced. But then not only were the injections skilfully performed, but the patients were kept quiet, avoiding all motion and compression. It is a good rule to perform the injections in the evening on those patients who are unable to remain at rest during the day. In hospital practice the patients do not make any objection to the numerous punctures sometimes required ; but in private practice the accompanying pain and subsequent inflammation are much less patiently borne. In most patients one injection was made per diem, and in several in two places, without any local inconvenience arising. But in some of them stomatitis was very quickly produced, without being attributable to any other cause. The number of injections has been very different, but when the treatment has been pursued uninterruptedly they have averaged between twenty-nine and thirty, carried over a space of five, and not infrequently six or seven weeks. The most unpleasant consequence observed has been the stomatitis, which in some cases has been very rapidly produced, sometimes even in six or seven days, and even quicker when the injection has been performed twice a day.

This is, indeed, most surprising, when we consider how little of the subli mate (often scarcely half a grain) has been introduced at a distant part. The mucous membrane of the mouth is alone affected, the salivary glands being little, if at all, concerned. As to the general result of his experiments with these injections, which, however, he acknowledges are at present insufficient in number, Prof. V. Sigmund considers they are an inferior means in the treatment of syphilis to the methodical mercurial inunctions which he has so long employed. Still in certain cases he regards injections as a valuable additional means of treating the disease. It is so in individuals who, from any cause, are unable to undergo inunction, and in those whose digestive organs are in a condition not to admit of their employing mercurials by the mouth. He has seen papular syphilis of young infants advantageously so treated, but they were children who were well fed and carefully looked after. He thinks great caution should be used with this means in patients suffering from kidney disease, as he has known such cases to become aggravated. Finally, all hygienic precautions are just as necessary in his mode of treating syphilis as in any other.

PROF. J. LEWIS SMITH, M. D., NEW YORK.

In infantile syphilis, the following formulæ may be employed :

1012. R. Hydrargyi cum cretâ,   gr.iij.-vj.
        Sacchari albi, $\ni$j.   M.
Divide into xij. powders ; one to be taken ter die.

1013. R. Hydrargyri chlor. corros.,   gr.j-ij.
        Syr. sarsaparillæ com.,   f.$\tilde{\mathfrak{z}}$ij.
        Aquæ,   f.$\tilde{\mathfrak{z}}$viij.   M.
A teaspoonful ter die.

Mercury, in whatever form employed, should not be discontinued entirely until several weeks after the syphilitic symptoms in the child have disappeared. It is proper to continue it for a time, in diminished quantity, after the health seems fully restored.

When the mecurial is omitted, tonics are often required. The preparations of cinchona are useful in these cases, as are also those of iron. The liquor ferri iodide is especially useful in this class of cases.

THOMAS HAWKES TANNER, M. D., F. L. S., LONDON.

> 1014. ℞. Hydrargyri chlor. corros., gr.ij.
> Pulveris opii,   gr.v–viij.
> Pulveris guaiaci,   ʒss.   M.
> Fiant pilulæ, xvj.

Once, twice, or three times a day, where it is desirable to continue the use of the corrosive sublimate over many weeks.

> 1015. ℞. Ammoniæ carbonatis,   ʒss.
> Potassii iodidi,   Əj.
> Tincturæ aconiti folii,   ℳ.xxx.
> Tincturæ cinchonæ flavæ, f.ʒvj.
> Aquæ menthæ piperitæ ad f.ʒiij.   M.

Tablespoonful in a half wineglass of water, ter die, at 9 A. M., 2 P. M, and 7 P. M.

> 1016. ℞. Hydrargyri iodidi viridis, gr.ij.
> Extracti opii,   gr.j.
> Extracti hyoscyami,   gr.vj.   M.

Divide into two pills, and order one to be taken every night at 11 o'clock, as long as the above mixture is continued. Very useful in many forms of constitutional syphilis.

EDWARD JOHN TILT, M. D., LONDON.

> 1017. ℞. Hydrargyri iodidi viridis, gr.j.
> Extracti hyoscyami,   gr.ij.   M.

An anti syphilitic pill, to be taken morning and night.*

---

*Haud Book of Uterine Therapeutics. Am. Ed. 1869. p. 334.

OPHTHALMIC THERAPEUTICS.

1. DISEASES OF THE CONJUNCTIVA.

GEORGE LAWSON, F. R. C. S., SURGEON TO THE ROYAL LON-
DON OPHTHALMIC HOSPITAL, MOORFIELD, ETC.

ACUTE CONJUNCTIVITIS.

In the treatment of *Acute Conjunctivitis*, (catarrhal ophthalmia), our author reccommends that, every two or three
hours, or oftener, if the case be a severe one, the eyes be
bathed with one of the following lotions, being careful at each
application to permit a small portion to flow into the eyes.

LOTIO ALUMINIS.

1018. R. Aluminis, gr.vj.
       Aquæ destillatæ, f.ℨj. M.

LOTIO ALUMINIS MITIOR.

1019. R. Aluminis, gr.iv.
       Aquæ destillatæ, f.ℨj. M.

LOTIO ALUMINIS CUM ZINCI SULPH.

1020. R. Aluminis, gr.iij.
       Zinci sulphatis, gr.j.
       Aquæ destillatæ, f.ℨj. M.

Cold water should be employed between the times of these
applications, to keep the eyes free from the discharge.

A solution of nitrate of silver (grs. 1 to 2 to the ounce), is
useful, particularly when there is chemosis of the conjunctiva
and swelling of the lids. Two or three drops of this should
be dropped into the eye twice a day; the eyes being kept clear
of discharge by bathing them in cold water as often as may
be necessary, A little *unguentum cetacei* should be smeared
along the tarsal borders of the lids at night to prevent their
agglutination.

A purgative should be administered at the beginning of the attack. If the patient be hot and thirsty, an alkaline or effervescing draught may be prescribed, such as

MISTURA POTASS.E CITRATIS.

1021. R. Potassæ bicarbonatis,   Ɖj.
           Spts ammoniæ aromataci,
           Tincturæ aurantii,   aa f.ʒss.
           Aquæ destillatæ,   f.ʒiss.   M.

To be taken in effervescence with acidi citrici, gr. xiv, dissolved in a tablespoonful of water.

The spiritus ammoniæ aromatici, may be omitted if desired.

As a rule, tonics, such as bark, quinine and iron, are indicated after the first febrile symptoms ushering in the attack have subsided.

CHRONIC OPHTHALMIA.

Our author recommends as local applications, when there is any extra secretion present, stimulating drops or lotions, such as what he terms his

GUTT.E ARGENTI NITRATIS.

1022. R. Argenti nitratis,   gr.j.
           Aquæ destillatæ,   f.ʒj.   M.

Or,

GUTT.E ZINCI SULPHATIS.

1023. R. Zinci sulphatis,   gr.j–ij.
           Aquæ,   f.ʒj.   M.

These solutions should be dropped into the eye twice a day.

Lotions with alum, or alum and zinc combined (F. 1019, 1020) are very efficacious.

If there be no abrasion of the cornea, the following lotion will be useful :

1024. R. Plumbi acetatis,   gr.ij.
           Acidi acetici diluti,   ℳ.ij.
           Aquæ destillatæ,   f.ʒj.   M.

At night, if there be much secretion from the Meibomian follicles, the tarsal edges of the lids should be annointed with

UNGUENTUM HYDRARGYRI NITRATIS DILUTUM.

1025. R. Ung. hydrarg. nitratis,   Ʒj.
           Unguenti cetacei,   ℥ij.   M.

Stimulating applications should not be made to the eye
when there is much photophobia, for they then fail to do good,
and are apt to act as irritants.

Counter-irritation is frequently beneficial in chronic ophthalmia. A small blister may be applied to the temple or
behind the ear, and repeated in two or three nights if necessary.

If the above remedies fail to afford relief, a seton of a single
or double thread of thick, corded silk inserted in the temple
will occasionally do good. It should not be allowed to remain
longer than three or four weeks, for fear of producing an un-
sightly scar.

In cases of persistent chronic ophthalmia, the lids should
be everted and carefully examined for granulations. If these
be present, the ophthalmia will continue until they are cured.

J. SOELBERG WELLS, PROF. OF OPHTHALMOLOGY, IN KING'S
COLLEGE, LONDON, ETC.

HYPERÆMIA OF THE CONJUNCTIVA

Is not unfrequently met with as a consequence of close application of the eyes to small objects by artificial light, or
from contact with atmospheric or mechanical irritants. The
cause is first to be removed. In order to relieve the feeling
of heaviness which oppresses the eyelids, employ one of the
following

EVAPORATING LOTIONS:

1026. R. Spiritûs ætheris nitrosi,   f.℥j.
           Acidi acetici aromatici,   gtt.vj.
           Aquæ destillatæ,   f.℥vj.   M.
To be sponged over the closed eyelids and around the eyes
   three or four times daily, and allowed to evaporate.

1027. R. Ætheris,   f.ʒij–iv.
           Spiritûs rosmarinæ,   f.ʒiv.   M.

To be used in the same manner as F. 1026, but in smaller
quantity, especially if the skin be delicate and suscepitible.

The best *astringent lotions* are the following :

1028. R. Zinci sulphatis,   gr.ij–iv.
           Aquæ destillatæ.   f.ʒiv–vj.   M.

1029. R. Plumbi acetatis,   gr.ij–iv.
           Aquæ destillatæ,   f.ʒiv–vj.   M.

The above are to be applied by saturating a piece of lint
with the solution, and laying it over the eyelids for 15 or
20 minutes, several times a day, allowing a few drops to
enter the eye.

In chronic cases of hyperæmia these applications must give
place to weak *collyria*, such as F. 1023, or

1030. R. Cupri sulphatis,   gr.j–ij
           Aquæ destillatæ,   f ʒj.   M.

1031. R. Argenti nitratis,   gr.i–ij.
           Aquæ destillatæ,   f.ʒj.   M.

A drop or two of one of these collyria is to be applied to
the conjunctiva. The sulphate of copper or lapis divinus may
be used in substance by touching the part lightly. The eye-
douche or cold compresses should follow each of these applications.

The popular error that it is beneficial to the eyes to dip the
face into cold water with the lids open is an injurious one, as
it often leads to, or aggravates, the affection under consideration.

THE EYE-DOUCHE.

The form of this instrument, recommended by our author,
is a piece of india-rubber tubing, about 4½ feet in length,
carrying a rose at one end, and at the other a curved piece
of metallic pipe, which is to be suspended in a jug of water
placed on a high shelf. The fine jet of water thrown up

through the rose should be about 12 or 15 inches in height; the force of it may be regulated by removing it from or approximating it to the eye. This form of eye-douche is much preferable to that applied by means of a cup, which is too strong and may increase the irritation it is intended to relieve.

The douche is to be employed night and morning, or oftener if the eyes feel hot and tired, for two or three minutes at a time. The eye-lids are to be closed and the stream directed gently against them.

The steam atomizer or the instrument used for ether spray, will also be found very useful and agreeable for the purposes of an eye-douche.

PROF. JOSEPH PANCOAST.

1032.   R.   Zinci acetatis,   ℨss.
             Acidi acetici diluti,   f.ℨss.
             Aquæ,   f.ℨviij.   M.
An astringent collyrium.

MEMBRANOUS AND DIPHTHERETIC CONJUNCTIVITIS.

F. A. POPE, M. D., NEW ORLEANS, IN CHARGE OF THE UNIVERSITY EYE AND EAR CLINIC.

When it is certain that a case is one of diphtheretis, that is, one in which there is infiltration of the conjunctiva, with diminished vascularity and tendency to the formation of false membranes, cauterization and the use of astringents are contra-indicated. Frequent *cleansing of the eye,* the application of *cold water dressings,* and the careful use of *mercurials,* are the principal means of treatment.

In the early stages of the disease, the *application of leeches* to the temple is often of decided advantage.

In a case of diphtheritic conjunctivæ, it is only when the second stage of the disease has arrived, namely, that of restored vascularity and commencement of purulent secretion, that the use of nitrate of silver can be resorted to. The third stage, or that of cicatrization, can be but little benefited by treatment.

The solution of nitrate of silver preferred by our author is of the strength of gr. vj. to the f.ʒj. In administering mercury he orders gr. 1-10 of calomel every two hours, and mercurial inunctions upon the temple three times a day, or mercurial inunctions alone, upon the temple and in the axilla, every two hours.*

GONORRHŒAL CONJUNCTIVITIS.

DR. ROGERS, OF MADISON, INDIANA.

1033. ℞.   Acidi carbolici,   gr.j.
           Atropiæ sulphatis,   gr.ss.
           Zinci sulphatis,   gr ij.
           Aquæ destillatæ,   f.ʒj.   M.

This solution is to be dropped into the eye every two hours, and applied constantly with moist compresses externally.

Our author has proved the efficiency of this treatment in numerous cases of gonorrhœal conjunctivitis, with chemosis, great swelling of the lids, profuse purulent discharge, photophobia, etc. A week originally suffices for a cure.†

PURULENT OPHTHALMIA.

PROF. GUNNING S. BEDFORD, NEW YORK.

1034. ℞.   Hydrarg. chlor. corros.,   gr.j.
           Ammoniæ muriatis,   gr.iv.
           Aquæ destillatæ,   f.ʒvj.   M.
       Ft. sol.

For purulent ophthalmia in new-born infants, the eyes to be washed with the solution several times during the day. The applications should not be confided to the nurse; they should be made by the practitioner himself, as follows: The child

---

*New Orleans Journal of Medicine. April, 1868, p. 286.

† Western Journal of Medicine, and the Medical Archives for July, 1869, p. 453.

being placed on its back, resting in the lap of the nurse, the practitioner placing its head on his knee, with a soft sponge, moistened with tepid water, cleanses the eyes. The lids are then gently separated, and after everting them, the accumulated matter is removed, and the collyria applied.

It may become necessary to touch the inflamed conjunctiva by means of a camel's-hair pencil, with the following solution once a day :

1035. R. Argenti nitratis,   gr.ij.
           Aquæ destillatæ,   f.ʒj.   M.
       Ft. sol.

When the child falls asleep, the outside borders of the lids, in order to prevent their agglutination, should be smeared wish fresh butter, fresh olive oil, or, what perhaps is better, the red precipitate ointment. The bowels are to be kept regular with castor oil, or flake manna in solution, and above all, the eyes are to be protected against light. This treatment, if faithfully carried out, will effect a cure, and should not be surrendered for leeches, blisters, etc.

Prof. Gross states that in the purulent ophthalmia of infancy he has usually effected excellent and even rapid cures by the injection, every few hours, of tepid water, or milk and water, followed immediately after by a solution of bichloride of mercury, from the eighth to the twelfth of a grain to the ounce of water, and the constant application of a light elm poultice, medicated with acetate of lead, and frequently renewed. The bichloride of mercury is of all local remedies in this affection the most efficacious in its action, making generally a most rapid and decided impression upon the discharge. Very weak solutions of lead, zinc and alum are also advantageous, but altogether inferior to the bichloride. One of the great points in the treatment of this and other forms of purulent

ophthalmia is to get rid of the acrid secretions, which, if allowed
to remain, always act as irritants. As to leeches and counter-
irritants, they should never be employed in this disease as it
occurs in infancy.

1036. R.	Pulveris opii,	gr.ij.	
	Hydrargyri,	gr.1–s.	
	Aquæ,	f.℥j.	M.

To be injected every half hour, in purulent, gonorrhœal,
and other forms of ophthalmia. If there be unusual swelling, and a rapid extension of the morbid action, the most
appropriate measures are, free incision of the outer surface of
the lids, extensive scarification of the chemosed conjunctiva,
together with the use of the above injection. If the discharge
of pus is very profuse, the inner surface of the lower lid may
be pencilled over twice a day with a strong solution of nitrate
of silver, (from the eighth of a grain to two grains to the
ounce for the more ordinary cases, while in the more violent,
it may range from five to sixty). When the solution is very
strong, it should be applied by means of a camel's-hair pencil,
the inflamed surface having been previously dried with a soft
linen rag. The solid nitrate of silver ought never to be used
about the eye. When the lids are enormously swollen, (in
the adult), great benefit is derived from the application to
them of a large blister, the surface being well protected with
gauze to prevent the fly from falling into the eye. The use
of the syringe is of paramount importance in these cases, as
it is the only means by which it is possible to obtain clearance
of the irritating matter, and effectually medicate the inflamed
surface.

SUB-ACUTE OPHTHALMIA.

PROF. JOHN B. BIDDLE.

1037. R. Zinci sulphatis,   gr.ij.
           Morphiæ sulphatis,   gr. 1–8.
           Aquæ rosæ,   f.$\frac{3}{3}$j.   M.

As an application for sub-acute ophthalmia.

WEAKNESS OF EYES.

PROF. ALFRED STILLÉ

Speaks of the vapor of the oil of rosemary, produced by rubbing a few drops between the palms of the hands, and then
allowed to come in contact with the eyes, as having been used
with advantage in weakness of these organs from nervous
exhaustion.

BURNS AND SCALDS OF EYES AND LIDS.

GEORGE LAWSON, F. R. C. S., ENG.

1038. R. Glycerinæ,
           Aquæ rosæ,   aa. f.$\frac{3}{3}$ij.
           Aquæ destillatæ   ad f.$\frac{3}{3}$viij.   M.

A soothing lotion for washing the eye and lids in cases of
burns and scalds. A few drops of olive oil should be dropped
into the eye, and the lids then gently closed, and some cotton
wool laid closely over them, which may be kept in its place
by a single turn of a light bandage. The dropping of the oil
into the eye should be repeated two or three times during the
day, and each time the bandage is removed the above lotion
should be employed to remove any discharge which may
have accumulated. This is the only treatment slight cases
require.

TREATMENT OF IRITIS.

JOHN HUGHES BENNETT, M. D., F. R. S. E., PROF. IN THE UNIVERSITY OF EDINBURGH.

Our author has recorded five cases of *rheumatic iritis* treated
without mercury. They all recovered. He employed atropia

locally, and quinine internally. In one case of double rheumatic iritis, with conjunctivitis, of the most severe description, recovery took place of the right eye in five weeks, and of the left in six weeks.

ROBERT BRUNDNELL CARTER, F. R. C. S.

1039. ℞.   Atropiæ sulphatis,   gr.ij-iv.
             Aquæ destillatæ,   f.ℨj.   M.

Our author employs this solution as a local remedy, in all cases of iritis, whatever its origin or constitutional cause.   It must be dropped into the eye, at first at short intervals, as every hour. When dilatation is produced, or when, after a few applications the pupil still resists dilatation, the application is to be continued two or three times a day.

Dr. HEYMANN, of Dresden, advises in severe cases the application of a particle of *solid atropia*, or of its sulphate, to the tarsal conjunctiva, as being more certain and powerful in its action than any solution ; in some instances our author has obtained good results from this practice.   He considers, however, that the precise method of application is of minor importance.   The point to be borne in mind is that the use of atropia in every case of iritis, whatever else may be done or left undone, is the one thing that should never be omitted at the outset of the treatment.

The chief value of atropia does not depend upon its power of dilating the pupil.   In the most severe cases of iritis it does not begin to dilate the pupil until the inflammation has first been in some degree subdued, or has subsided—and in such cases its influence is more marked and beneficial than in others.   One author explains this influence in few words, by saying that, besides diminishing hyperæmia, by producing some contraction of the blood-vessel, it secures rest to the parts within the eye by paralyzing the muscles of accommodation.

In the majority of obstinate cases of iritis there is pain
enough to constitute a marked feature of the disease. As long
as there is pain there will be no improvement; and this pain
is, commonly, merely a symptom of the persistence of the
cause of the nervous irritation in which iritis has its origin.
Our author holds it to be a principle that the pain of iritis
must always be subdued by anodynes; not merely mitigated,
but absolutely mastered. If there be no pain, no anodyne
will be needed. If there be only "uneasiness," a moderate
dose at bed-time may be sufficient. If pain be severe, opium,
or some of its preparations, should be measured only by their
effects—given hour after hour, until the pain is no longer felt,
and then continued at whatever intervals may be sufficient to
keep it in abeyance. Mr. Zachariah Laurence has published
reports of several cases of iritis treated successfully by opium
pushed almost to narcotism; but the secret of his success was
simply the removal of pain, and this result (for which small
doses will often suffice), is both the explanation of the *modus
operandi* and the test of the quantity that should be administered. The preparation employed is a matter of little consequence. For the sake of rapidity of action, it is often well to
commence by injecting a full dose of morphia under the skin
of the temple; and pills of soft opium afford a manageable
means of continuing the effect.

When all hitherto described has been carried out, there
will still remain cases in which, notwithstanding the use of
atropia, the relief of congestion, and the subjugation of pain,
the pupil does not dilate, and vision either deteriorates or at
least does not improve. In these cases it is found as a mere
matter of fact, that mercury, given rapidly, but discreetly,
until the gums show some slight sign of its constitutional effect,
immediately break the chain of morbid action. From the
22

very day on which the mercurial line becomes apparent, the sensations of the patient are relieved, and the symptoms of inflammation decline.

He is accustomed to use the following :

1040. ℞. Pilulæ hydrargyri,   gr. iij.
          Pulveris opii,   gr. ss.   M.
For one pill ter die, for one, two or three days, according to the strength and condition of the patient.

He then orders diminishing doses until the gums show signs of action. Finally he directs one small dose daily until the condition of the eye is so much improved as to render relapse improbable.

He does not believe that all the good effects of mercury on iritis can be produced unless the line on the gums can be obtained. But the condition of "salivation" should never be brought about designedly.

During the whole period of treatment the eye should be closed and protected by a compressive bandage, applied with comfortable tightness over a pad of jeweler's cotton wool. By this means patient will be enabled to walk abroad without restraint, so long as he avoids injurious fatigue or hurry. Sometimes, especially when resting quietly at home, a poultice will be a pleasant substitute for the pad and bandage; but neither the one nor the other should be applied until a quarter of an hour after the installation of the atropia, lest the solution should be absorbed and removed from the eye.

When the inflammatory symptoms are rapidly subsiding, the mercury, and probably the opium, may be entirely laid aside. But the continued use of atropia is necessary in order to prevent relapse ; and the pupil should be kept fully dilated until the eye is quite well. As long as the pupil is dilated the eye does not participate in the functional changes of its fellow, to which, therefore, moderate use may be permitted. An attack of any severity usually leaves behind a temporary proneness to conjunctival irritation, which the atropia may often assist to keep up. For this the cautious use of a mild astringent, such as

1041. R. Zinci sulphatis,   gr. iv.
            Aquæ destillatæ,   f. ℥iv.   M.

This collyrium will usually be found effectual.

It will often be desirale to protect the eye from the glare, wind, and dust, after a severe attack, by the use of blue glasses. These are now made of a watch glass form for the purpose of excluding side light.

CULLERIER (Surgeon to Hôpital du Midi, Paris), says that in syphilitic iritis, mercurial or saline purgatives, inunction, with mercurial ointment combined with belladonna, collyria of belladonna or atropia, and lotions containing opium to sooth the pain, in addition to the internal treatment suited to secondary symptoms, will suffice to effect a cure. He has never discovered any reason for sounding the praises of turpentine, so highly recommended by some authorities. Prof. Gross also reports adversely upon the use of turpentine in this affection.

GEORGE GASCOYNE, F. R. C. S. (Vol. LII. Medico-Chirurgical Transactions), reports eighteen cases of *syphilitic iritis* treated without mercury. He asserts that the iritis which occurs in syphilis is not only amenable to a simple local treatment, but that the results are fully as favorable as when mercury is used. In all his recent cases the eye completely recovered. In those in which the iris had contracted adhesions before local treatment was adopted, perfect vision was regained in most, and useful vision in all. The average time during which the atropine drops were continued was about twenty-six days; the shortest period being fourteen and the longest forty-nine. He keeps the eye shaded for several days after the complete disappearance of the lymph and the return of the natural color.

A number of years ago Dr. H. W. WILLIAMS, of Boston, recorded (in the Boston *Medical and Surgical Journal*) sixty-four cases of iritis treated without mercury. The cases include every degree and variety of the affection, the idiopathic, rheumatic, syphilitic, and traumatic forms. In all excepting four a good recovery was obtained. In all the four unfavorable cases, the disease had been neglected in its early stages.

GEORGE LAWSON, F. R. C. S., SURGEON TO THE ROYAL LONDON OPHTHALMIC HOSPITAL, MANSFIELD.

In the treatment of *syphilitic iritis* our author regards mercury as imperatively called for. It should be given in doses sufficiently large and frequent to bring the patient quickly under its influence, but as soon as the gums begin to grow tender and spongy, the quantity should be diminished so as to avoid anything like profuse salivation. A piece of the size of a nut of the *unguentum hydrargyri* may be rubbed into the axilla night and morning, or a pill with calomel and opium may be administered:

1042. R. Hydrarg. chloridi mitis,   gr.j–ij.
      Pulveris opii,   gr.¼–½.
      Confectionis rosæ,   q. s.   M.
   For one pill, ter die.

If the patient be feeble, quinine may be prescribed at the same time, and they may be conveniently ordered in the following mixture:

1043. R. Quiniæ sulphatis, gr.xij.
Acidi sulphurici diluti, f.ʒij.
Tincturæ aurantii, f.ʒvj.
Aquæ destillatæ, q. s. ad f.ʒvj. M.

Tablespoonful in water, ter die, while the mercurial inunction
is used night and morning.

If the patient has already been salivated before he first
comes under treatment, the following iodide of potassium
mixture should be given:

1044. R. Potssii iodidi, gr.xxxvj.
Potassæ bicarb., ʒj.
Infusi quassiæ, ..ʒvj. M.

A tablespoonful ter die.

At the same time a slight mercurial action may be kept
up by the use of the following:

UNGUENTI HYDRAGYRI CUM BELLADONNA.

1045. R. Extracti belladonnæ, ʒj.
Unguenti hydrargyri, ʒvij. M.

To be rubbed into the brow and temple, and allowed to remain on during the day.

When all the effused lymph has been absorbed and the
iritis has nearly subsided, the murcurial medicines should be
omitted, but the iodide of potassium should be continued for
two or three months combined with a bitter tonic, or if the
patient is anæmic, with some preparation of iron, as the

MISTURA POTASSII IODIDI CUM FERRO.

1046. R. Potassii iodidi, gr.xxxvj.
Potassæ bicarb.,
Ferri et ammoniæ cit., aa. ʒj.
Aquæ destillatæ, f.ʒvj. M.

A tablespoonful in water ter die.

If the iritis recur after some months, or if it assume a
chronic form, the following mixture will be found of great
service:

1047. R.   Hydrarg. chlor. corros.,   gr.j.
            Potassii iodidi,   ℨj.
            Tincturæ columbæ,   f.ℨij.
            Aquæ destillatæ,   q. s. ad f.ℨvj.   M.

Two teaspoonfuls in a glass of water two or three times a day.

Atropia is essential in the treatment of every form of iritis, and should be ordered at the very commencement of the attack, and persevered in during its continuance. A solution of the strength of gr. ij., to aquæ f.ℨj. should be dropped into the eye two or three times a day. When the atropia fails to give ease, or acts, as is sometimes the case, as an irritant, the following belladonna lotion may be employed:

LOTIO BELLADONNÆ.

1048. R.   Extracti belladonnæ,   ℈ij.
            Aquæ destillatæ,   f.ℨviij.   M.

Rheumatic iritis does not require the active mercurial treatment recommended for the syphilitic form of the disease. F. 1044 may be given during the day, and at night the following pill:

1049. R.   Hydrarg. chloridi mitis,   gr.j.
            Pulv. ipecacuanhæ comp.,   gr. v.   M.
       For one pill.

Or the mercurial and belladonna ointment (F. 1045) may be rubbed daily into the temple.

In some cases the treatment will fail to give relief. Then *quinine* in two grain doses may be ordered with benefit. Or, the quinine may be combined as follows:

1050. R.   Quiniæ sulphatis,   gr. xij.
            Tincturæ ferri chloridi,
            Acidi nitrici diluti,   aa. f. ℨj.
            Aquæ destillatæ,   f. ℨvj.   M.

A tablespoonful in water, to be taken through a tube, ter die.

When there is great photophobia and pain in the eye, the quinine, or quinine and iron treatment, together with a mild

mercurial inunction into the temple, will be found most useful. To relieve the pain a fourth or a third of a grain of the
acetate of morphia may be injected subcutaneously into the
arm. Our author directs the following formula for the

INJECTIO MORPHILE.

1051. R. Morphiæ acetatis,   ℈iv.
          Aquæ destillatæ,   f. ℥j.   M.

Rub the morphia gradually with the water and add a few
drops of dilute acetic acid if necessary for perfect solution.

♏. vi–gr. j. of acetate of morphia.

Turpentine has been prescribed with advantage in obstinate
cases of *non-syphilitic iritis.* It may be ordered as follows:

1052. R. Olei terebinthinæ,   f. ℥iij.
          Syrupi acaciæ,   f. ℥iss.
          Aquæ pimentæ,   f. ℥iv.   M.

A teaspoonful four or five times a day.

During the whole time the pupil should be kept well dilated
by means of atropia, or the belladonna lotion (F. 1048).

N. C. MACNAMARA, PROF. OF OPHTHALMIC MEDICINE, CAL-
CUTTA.

1053. R. Atropiæ,   gr.iv.
          Aquæ,   f.℥j.   M.

To be dropped into the eye three times a day, in cases of
syphilitic iritis in children.

Mercurial ointment should also be rubbed into the thighs
every other night, for twenty minutes; and thirty drops of
cod-liver oil, with one half a grain of iodide of iron, should be
administered twice a day to an infant six months old. For
syphilitic iritis, mercury judiciously employed is the sheet
anchor to be relied upon. The best mode of employing it in
these cases is by inunction. Our author never prescribes
mercury internally for children, nor does he find it necessary
to push the treatment so far as to affect the gum.

According to MACKENZIE, and indeed all the best authorities, atropia ought to be employed as a collyrium in every case of iritis, and in all stages of the disease, to prevent unnatural adhesions of the iris. The earlier in the affection the remedy is administered the better.

Prof. GROSS says that the best remedy for iritis, in all cases, excepting perhaps the most simple, is mercury, carried to the extent of rapid ptyalism. He states that atropia does not possess the property of dilating the pupil in iritis, for the moment the iris is actively inflamed, it ceases to be influenced by narcotic applications; the pupil contracts, and no stimulus, however powerful, can afterwards excite it. In the syphilitic variety of the complaint, iodide of potassium may be advantageously exhibited, to aid in completing the cure, after having made a fair trial of mercury.

DISEASES PECULIAR TO WOMEN.

TREATMENT OF AMENORRHŒA.

J. M. DA COSTA, M. D.

1054. ℞. Apiol,   gr.iv.

In the form of a granule or "pearl," four times a day as an emmenagogue. To be taken for three days before the expected period. Apiol is an excellent remedy for amenorrhœa when there is no uterine disease. It is also useful in dysmenorrhœa.

Dr. TILT says that given every two hours, so soon as the pains of dysmenorrhœa begin, it acts like a charm in some cases of nervous dysmenorrhœa, but it is of little use when the dysmenorrhœa depends upon some disease of the womb.

EMMENAGOGUE PILL.

C. W. FRISBIE, M. D., EAST SPRINGFIELD, N. Y.

1055. R .   Assafœtidæ,
          Myrrhæ,   aa. ℥j.
          Aloes soccatrinæ,   Ɖij.
          Ferri lactatis,   ℥j.   M.

For lx pills. One night and morning.

T. HAWKES TANNER, M. D., F. L. S., LONDON.

1056. R .   Potassii bromidi,   ℥j.
          Tinct. cantharidis,   f. ℥iss.
          Tinct. cinnamomi,   f. ℥vj.
          Aquæ,   q. s. ad f.℥iss.   M.

Dessertspoonful three times a day, as a stimulating emmenagogue.

MENORRHAGIA.

T. HAWKES TANNER, M. D., F. L. S., LONDON.

1057. R .   Acidi gallici,   gr. xv.–xxv.
          Acidi sulph. arom.,   ℳ. xv.–xx.
          Tinct. cinnamomi,   f. ℥ij.
          Aquæ destillatæ,   q. s. ad f.℥ss.   M.

For one dose.

Mix with two or three tablespoonsful of water, and take every few hours in profuse menorrhagia, until the bleeding ceases.

Professor T. GAILLARD THOMAS, New York, says that in a case of menorrhagia the patient should be kept perfectly quiet upon her back; clothes wrung out of cold water should be laid over the uterus, vulva, and thighs; cold, acidulated drinks should be given freely; and the injection of all warm fluids strictly interdicted. In addition, the apartments should be kept cool, the nervous system quiet by opium or an appropriate substitute, and all conversation prohibited. In mild cases this may suffice, but in severe ones it will not. Then the speculum should be introduced, a sponge-tent passed into the cervix, and the vagina filled with a tampon. This will rarely fail. But in certain cases, as, for instance, those of cancer of the neck, the tent will not be admissible. Under these circumstances, a soft sponge or wad of cotton should be saturated with a solution of tersulphate of iron, laid upon the cervix, and the tampon placed against it, or a small linen bag may be filled with powdered alum, placed in contact with the cervix, and held in place by a tampon; or two drachms of tannin may be left free against the part. To these means almost all cases will temporarily yield, more especially if the use of the tent be admissable.

EMMENAGOGUE ENEMA.

EDWARD JOHN TILT, M. D., M. R. C. P., LOND.

1058. R. Aloës Barbadoensis,   gr.x.
           Tepid milk,   f.ℨiij.   M.

To be injected twice a day when the menstrual flow is due, until it comes, or until tenesmus becomes unbearable.

EMMENAGOGUE VAGINAL SUPPOSITORY.

1059. R. Aloin,   gr.ij.
           Buttyri cocoæ,   gr.x.   M.

EDWARD JOHN TILT, M. D., M. R. C. P., LONDON.

1060. R. Olei terebinthinæ,   f.ℨss.
           Tincturæ capsici,   f.ℨss.
           Tincturæ ergotæ,   f.ℨj.
           Tinct. lavend. comp.,   f.ℨij.   M.

In cases of uterine hemorrhage, give from half a drachm to a drachm of this mixture in milk, after shaking the botttle. In severe flooding after parturition, from half an ounce to an ounce may be given in plenty of milk, with good results.

PROF. ELLERSLIE WALLACE, PHILADELPHIA.

1061. R. Aloes,   gr.v.
           Olei tanaceti,   gtt.xl.
           Cantharidis,   gr.vij.
           Ferri lactatis,   ℈iv.   M.
       Fiat massa; in pilulas xxviij. dividenda.
   One, morning, noon, evening and night, as an emmena-
       gogue.

DR. RUBEN, OF HAMBURG.

1062. R. Ergotinæ,   gr. xv.
           Glycerinæ,
           Aquæ destillatæ,   aa. f. ʒss.   M.
   Dose, fifteen minims.

Our author has used ergotin in severe cases of menorrhagia
with good results.   In one case the hemorrhage had continued
for four months at the time the patient came under treatment.

ACCIDENTAL HEMORRHAGE DURING PREG-NANCY.

J. G. SWAYNE, M. D., PHYSICIAN ACCOUCHEUR TO THE BRISTOL
GENERAL HOSPITAL, ETC., ENGLAND.

The following formulæ are of service in cases in which the
hemorrhage occurs before full term:

1063. R. Acidi sulphurici dil.,   f.ʒj.
           Tincturæ opii,   m.xl.
           Infusi rosæ comp.,   f.ʒvj.   M.
       Two tablespoonfuls every other hour.

1064. R. Plumbi acetatis,   gr.xviij.
           Acidi acetici,   m.xx.
           Morphiæ acetatis,   gr.j.
           Aquæ destillatæ,   f.ʒvj.   M.
       Two tablespoonfuls every hour.

The woman is also, of course, to be kept in a recumbent
position, and cold compresses applied to the abdomen and

vulva. Cold drinks and cold water enemata may be administered. By the employment of these expedients the bleeding may be checked and the patient carried in safety to the close of her pregnancy.

DYSMENORRHŒA.

THEO. JEWETT, M. D., PROFESSOR OF OBSTETRICS, BOWDOIN MEDICAL COLLEGE.

1065. ℞. Camphoræ,   ℥ijss.
          Extracti belladonna,
          Quiniæ sulphatis,   aa. ℈ss.
          Pulveris acaciæ,   q. s.   M.
      For lxxx pills.
  One to be taken every four hours until relieved.

1066. ℞. Ext. scutellariæ fluidi,
          Decocti aloës compositi, aa. f. ℥ss.   M.
  A desertspoonful every two or three hours until relieved.

Dr. C. W. Frisbie, of East Springfield, N. Y., writes that he used the above formulæ in his practice many times, and, when the cases have been properly selected, with the most happy results.

INJECTIONS OF WARM WATER IN THE TREATMENT OF UTERINE INFLAMMATION AND DYSMENORRHŒA.

DR. A. DESPRES, SURGEON TO THE LOURCINE HOSPITAL, PARIS.

Our author states that injections of hot water of from 95° to 104° F. are excellently calmative and powerfully antiphlogistic in the treatment of uterine inflammation, and that when they provoke a sanguineous discharge, it is a forerunner of improvement.

When there is a periuterine inflammation, even about a hæmatocele, warm water is still a good resolvent, and hitherto he has not seen that warm water augmented the hemorrhage. It is true to say that the vaginal injection never reaches the vessel that emits the blood.

In dysmenorrhœa warm water occasions congestion of the uterus, and the congestion is followed by a return of the menses, and consequently by a marked alleviation.

Finally, injections of warm water act like the cataplasm and warm lotions, which are so usefully employed in inflammation of the integument.

The injections of warm water are practiced at the hospital with irrigators, of which the jet is not very strong. The water used should be of 95° to 104° F., and it is renewed two, four or six times in the day. This therapeutic means is convenient and not repugnant to the patient—a good condition for its employment; besides, it occasions no bad result.

THE CONSTITUTIONAL TREATMENT OF UTERINE DISEASES.

HENRY M. FIELD, M. D., OF BOSTON.

Our author lays a great stress upon the necessity of associating constitutional medication with topical applications in the treatment of uterine disease. For, firstly and theoretically, the uterine lesion, if it have existed for sometime before it is brought to our notice, although it was the original cause, is not at present, the sole and efficient cause of the patient's condition, but also the depraved state of the blood, and of the nervous system, and the many forms of functional derangement which complicate the case; and moreover, secondly and practically, we cannot hope successfully to compete with even a local inflammation, or to restore a single diseased organ to a normal and healthful condition, so long as the blood is seriously impaired in its quality, and the more important functions of animal life are depraved or disordered.

Nevertheless, he suspects that there is a too general tendency, in uterine therapeusis, to trust solely, or nearly so, to the employment of local medication. This was certainly, at one time, a fault of his own, but his experience has more recently included cases which have abundantly satisfied him of the frequent fallacy and ineffectiveness of such practice.

He has seldom found *complicating constipation* cured by the removal, for instance, of the mechanical obstacle imposed upon the bowel by the flexed and engorged uterus. Its more or less long continuance has produced too profound an impression, and extended in its influence, to the procuring of other abnormal conditions of the economy, which are, in their turn by acting reflexly, concerned in keeping it up, for it to be permanently and effectually removed by the removal only of the first, although it be the principal, cause, in the series of causes which produced it. Many of these cases, of course, are essentially cases of insufficient innervation, and the use of nux vomica or strychnia is indicated; but, at times, the entire system is so perverted or overwhelmed by the reaction upon it of the uterine disease, that a therapeusis, as broad in its application as are the indications of a state of universal ill-health, will be required before we can expect the return of health and regularity to so important a function as that of the bowels. Such are those cases in which the nervous system has especially suffered from the effects of the uterine lesion : and such, even more markedly, in which the quality of the blood has become very much depraved. With such patients to give strychnia or belladonna, with the design of acting specifically upon a single function, and of restorting a condi-

tion of permanent health to the bowels, would be almost as
short-sighted, and almost as much of a temporary expedient
as it would be to give purgatives. The impoverished blood
must first be fortified and enriched before we can look for the
normal performance of any important function; and accordingly there are patients, answering the condition described, in
whom a course of iron, properly regulated, may be the only
general therapeutic agent that is necessary to overcome and
cure constipation, which could not be reached in any other
way.

In some of his uterine cases he has found arsenic, and especially the arseniate of iron, very effective in removing constipation, and has sometimes received benefit from this agent
when he had failed to make a successful impression with any
of the more commonly used remedies.

He has been very much pleased in cases of female difficul-
ties in which iron has been indicated with the action of *oxalate of iron*, a preparation first brought to his notice by Prof.
CRAIG, of the Smithsonian Institute. Being a light and taste-
less powder, with nothing repulsive in its appearance, it can
be exhibited in that form to those occasional patients who
are unable to swallow a pill. He values it particularly
cause it is less liable to cause irritation or derangement of the
stomach, or constipation of the bowels, where this common
effect of ferruginous preparations is to be avoided, than is any
other form of iron with which he is familiar. He has found
it to agree with and benefit patients who from past experience, believed themselves unable to take iron in any form.

He urges the employment of conjoined constitutional medication in the treatment of uterine diseases, as required, for
two principal reasons: First, on account of disturbances or
derangements of special functions, with, or without, a state of

general ill-health on the part of the patient ; and, second, for the favorable reaction of such medication upon the womb itself.

Syringes are so extensively used in medicine and surgery that an improved instrument may be considered a boon to the profession.   Those in ordinary use are of threefold construction.

1. Those which act by direct pressure upon the fluid through the medium of a piston and plug.

2. Those which act by exhausting the air, and afterward expelling the fluid which enters into the vacuum.

3. Those which act like SCANZONI's, on the principle of a syphon.

All these appliances are liable to get out of order, and neither of them is perfect.   Dr. BEIGEL, has combined all their separate advantages in one instrument.   It consists of a glass bottle holding the fluid to be injected.   Into the neck of the bottle a cork or plug is fitted.   Through this cork a tube descends into the fluid, and at right angles to it two other tubes are affixed.   One is in communication with a hand-ball bellows whilst the other has a perforated end and is inserted into the vagina.   On using the hand-bellows, a current of air is forced into the fluid contained in the bottle, which, compressing it, causes it to escape with some violence through the vaginal tube.   It works with precision, and is also applicable to the rectum, eyes, pharynx, nose, etc.

HERR D. TOLDT, has devised a somewhat similar syringe, but instead of a hand-ball or bellows, he employs the descent of a body of mercury for compressing the air and forcing the fluid outward.   The principle is the same in both.

CARBOLIC ACID IN ULCERATION OF THE OS UTERI.

DR. ROE, COOMBE LYING-IN HOSPITAL, DUBLIN.

Dr. Roe has been for some years in the habit of using carbolic acid as a local application in cases of ulceration of the os and cervix uteri, and has found it to yield results superior to any other topical treatment which he has tried. He has used it in cases where the whole round of other applications has been unsuccessful, and always with the most happy results. He agrees with Dr. ROBERTS, of Manchester, who last year drew the attention of the profession to the subject, in considering it a caustic, which, as regards its severity, may take intermediate rank between the nitrate of silver and strong nitric acid, besides acting as a disinfectant, a matter of no small importance in these cases. Dr. ROE does not use it in as strong a form as Dr. ROBERTS, and does not consider the strong acid necessary in very superficial ulceration. A mixture of one part of the strong acid with two of olive oil seems to answer all ordinary purposes; but in cases of very deep ulceration the use of the strong acid may be called for. In such cases Dr. ROBERTS desires the acid to be liquified by the addition of a very small quantity of water. This has not been found to answer the purpose in the Coombe Hospital, but it has been there discovered by Mr. WEIR, that the addition of a few grains of camphor will dissolve the acid, and will, moreover, prevent it again becoming solidified, even at a freezing temperature. The application of the carbolic oil to the os uteri is best affected by soaking a little cotton wool in the liquid, securing it by a string, and introducing it through a speculum, the string being left depending out of the vagina, and the patient being directed to pull it away on the second day. This procedure is repeated in ordinary cases about

23

twice every week. If it be desired to apply the acid to the cervical canal, it may readily be done, by passing in a gum elastic catheter smeared with the carbolic oil.

VULVITIS.

T. GAILLARD THOMAS, M. D., PROFESSOR COLLEGE OF PHY-
SICIANS AND SURGEONS, N. Y.

In the treatment of *purulent vulvitis*, if the inflammatory action run high, the woman should be kept in bed and upon a low diet. Saline cathartics should be administered. Cleanliness is to be carefully enjoined, and cooling emollient applications applied and retained upon the part. The vulva should be freely bathed three or four times a day with warm water, and a warm poultice of powdered linseed, slippery elm, or grated potato, with the addition of lead and opium, directed.

So soon as the acute symptoms have subsided, the following lotion should be kept in contact with the parts, by dossils of lint soaked in it and placed between the labia.

1067. R. Tincturæ opii, f. $\bar{3}$ij.
Plumbi acetatis, $\bar{3}$j.
Aquæ, Oj. M.

At a still later period the diseased surface should be painted over several times a day with

1068. R. Liq. ferri sulphatis,
Glycerinæ, aa. f. $\bar{3}$ss. M.

If this treatment be not effectual in eradicating the trouble, a solution of nitrate of silver (gr. x to aquæ f. $\bar{3}$j) should be applied by means of a brush every other day, and the part kept constantly powdered with lycopodium, bismuth, or starch, until the recovery is completed.

PRURITUS VULVÆ.

A. C. GARRATT, M. D., BOSTON.

1069. R. Acidi hydrocyanici
(Scheele's), f. ℨij.
Liq. plumbi subacetatis. f. ℨiv.
Aquæ, f. ℥iij. M.
As a local application.

Dr. HORATIO R. STORER states that he has long given great comfort in this affection by Oldham's ointment of hydrocyanic acid and acetate of lead, with cocoa butter.

Dr. G. S. JONES, of Boston, has employed with benefit in pruritus of the vulva, the following :

1070. R. Sodæ biboratis, ℨj.
Camphoræ, ℨj.
Olei gaultheriæ, gtt. xxx.
Aquæ bullientis, Oij. M.
When cool, pass through a cloth. To be used cold as a wash for the parts, and as an injection into the vagina.

DR. LÉON GROS, PARIS.

Our author records a case in the *Bulletin Générale de Thérapeutique*, of a woman tormented during two successive pregnancies by a pruritus of the whole cutaneous surface without eruption. Nervous spasms caused by the itching rendered the woman's life miserable. Various treatments were tried without effect. Pyrosis and dental neuralgia at length complicated the case. At this period, *smoking of tobacco* was resorted to with complete and speedy success on both occasions. One cigar was smoked every night. Sleep and comfort returned.

OVARIAN NEURALGIA.

J. WARING CURRAN, L. K. AND Q. C. P. I., ETC.

1071. R. Ammoniæ muriatis, ℨij.
Tincturæ aconiti, f. ℨij.
Syrupi aurantii corticis, f. ℥viij. M.
A teaspoonful ter die in the treamtent of *ovarian neuralgia*.

Our author states that this combination has almost a magical influence in many cases.   He reports (*Medical Press and Circular*, August 19th, 1868,) six cases in which various sedatives and anodynes had been tried in vain.   In all he found that before the above mixture was finished by the patient the pains had entirely ceased.

Dr. T. J. Newman, of Chicago, confirms the usefulness of this mixture, and records (in the *Chicago Medical Examiner*, for November, 1869,) three cases of neuralgia of the ovaries treated by it with success, after the failure of other remedies.

INFANTILE THERAPEUTICS.

DIARRHŒAL AFFECTIONS.

JAMES S. HAWLEY, M. D., GREEN POINT, LONG ISLAND, N. Y.

In infantile diarrhœa the indications are as follows : First, to remove all sources of irritation from the quantity or quality of the ingesta, or change of temperature.   Second, to allay irritation by sedatives, of which the best are the preparations of opium and the salts of bismuth.   When irritation without pain exists, bismuth most promptly and satisfactorily allays it, but when accompanied with pain, the addition of a minute portion of opium becomes a necessary complement to its effectiveness.   Thirdly, artificial digestion by the administration of *pepsin.*

> 1072.   R·   Pulv. pepsinæ Americanæ,
> Bismuthi subnitratis,   aa. ʒj.   M.
> For x powders.
> One to be given every three or four hours to a child a year old.

Opium may be combined as follows:

1073. R. Pulv. pepsinæ Americanæ,
            Bismuthi subnitratis,   aa. ʒj.
            Pulveris opii,   gr. j.   M.
        For xij powders.

One to be given every two or four hours, according to circumstances.

THOMAS HAY, M. D., PHILADELPHIA.

Our author has employed the following treatment in cases
of cholera infantum with the best results:

1074. R. Hydrarg. chlor. mitis,   gr. ij.
            Bismuthi subcarbonatis,   gr. xvi–xl.
            Pulv. ipecacuanha comp.,   gr. j–ij.
            Pulv. sacchari albi,   gr. xij.   M.
        For viij powders.

One to be taken every three hours for two or three days, or
until the tongue and mouth become moist and the alvine
excretion changed in color and consistency.

Then the following powders are given, and will ordinarily
complete the cure:

1075. R. Bismuthi subcarbonatis,   gr. xvj–xl.
            Pulv. ipecacuanhæ comp.,   gr. j–ij.
            Pulv. aromatici,   gr. viij–xvj.
            Pulv. sacchari albi,   gr. xij.   M.
        For viij powders.

One to be taken every three or four hours, in the mother's or
cows's milk.

Counter-irritation is kept up over the abdomen with mustard plasters applied at intervals of three or four hours. The
infant is allowed to suck at a piece of ice held in its mouth.
When stimulants are required, the doctor gives from fifteen
to thirty drops of Port wine. When the infant is artificially
fed he gives it cow's milk and lime-water in the proportion of
one fluid ounce of the latter to five fluid ounces of the former;
also broiled mutton or beef minced very fine. All farinaceous
food is forbidden. The child must be nursed or fed at regular intervals, and not allowed too much at a time.

DRS. MEIGS AND PEPPER, OF PHILADELPHIA.

Our authors recommend in the treatment of *simple diarrhœa* of childhood sulphate of magnesia combined with laudanum, as follows:

1076.   R.   Magnesiæ sulphatis,   ℨj.
       Tincturæ opii deodoratæ,   gtt. xij.
       Syrupi simplicis,   f. ℥ss.
       Aquæ menthæ,   f. ℥ijss.   M.

Dose—At one or two years, a teaspoonful every two or three hours. For older children, the proportion of magnesia and laudanum should be doubled.

If this fails, recourse must be had to an astringent. The officinal mistura cretæ may be given in teaspoonful doses after each loose evacuation, three or four times a day, or tincture of krameria, may be added, thus:

1077.   R.   Tincturæ krameriæ,   f. ℨj–ij.
       Misturæ cretæ,   f. ℥ij.   M.
Dose—Teaspoonful repeated as above directed.

Powdered *crabs' eyes* will sometimes succeed after the failure of the chalk mixture. Our authors employ the following formula:

1078.   R.   Pulveris oculi cancrorum,   ℨj.
       Pulveris acaciæ,   ℥ij.
       Sacchari albi,   ℈j.
       Aquæ cinnamomi,
       Aquæ,   aa. f. ℥iss.   M.
A teaspoonful to be given four, five or six times a day.

M. BOUCHUT recommends the following prescription of this remedy employed by HUFELAND:

1079.   R.   Pulveris oculi cancror.,   gr. x.
       Syrupi rhei,
       Aquæ fœniculi,   aa. f. ℥ss.   M.
Dose—A teaspoonful every hour.

Our authors have also employed with advantage either alone or with F. 1077 and 1078, an *aromatic syrup of galls*, prepared as follows:

1080. ℞. Pulveris gallæ opt., ℥ss.
       Pulveris cinnamomi, ℥ij.
       Pulveris zingiberis, ℥ss.
       Spiritûs vini gallici opt., Oss.   M.

Let the ingredients stand in a warm place for two hours, and then burn off the brandy, holding some lumps of sugar in the flames. Strain through blotting paper.

Dose—15 to 40 drops, three or four times a day, or when the discharges are very frequent, every two or three hours.

In the chronic form of simple diarrhœa our authors have found of late years the following tonic very useful:

1081. ℞. Tincturæ nucis vomicæ, f. ℥ss.
       Tinct. gentianæ comp., f. ℥iij.
       Syrupi simplicis, f. ℥v.
       Aquæ, f. ℥ij.   M.

Dose—A teaspoonful three times a day, after meals, for children of three or four years of age.

Wine of pepsine is also efficacious in such cases, in doses of half a teaspoonful ter die.

J. LEWIS SMITH, M. D., PROFESSOR IN BELLEVUE HOSPITAL MEDICAL COLLEGE, NEW YORK.

Prompt measures are required in cholera infantum, as the child rapidly sinks under the prostrating influence of the frequent watery discharges. Some evacuant is indicated at the outset, if there be any irritating material in the stomach or bowels, causing or keeping up the trouble. Small doses of ipecacuanha (from two to five grains) are often beneficial. When, however, the stomach is very irritable and the alvine discharges fail to carry off the intestinal contents, calomel is the great remedy. As it is slow in its operation, castor oil may be administered after it with benefit, or its operation may be aided by a simple enema. It should not be given to the extent of more than one or two doses.

Our author thinks that unless the stomach is quite irritable, castor oil, syrup of rhubarb, or if there be acidity present, rhubarb and magnesia will generally be sufficient to remove the indigestible matter.

If there be no indigestible substance in the intestines, purgatives are contraindicated, as they are then hurtful. The continuance of the diarrhœa for several hours affords a pretty sure evidence of the removal of any irritating matter which may have been present, and hence no purgative is required. The objects of treatment then should be to diminish the frequency of the evacuations, and improve their character. No time should be lost. Opium in some form is the chief reliance.

If laudanum be used, it may be administered in one drop doses, every two or three hours to a child one year old. Its effect should be watched. If the evacuations are partially checked, *and there are signs of stupor,* stop the opiate, or at least give it less frequently.

Astringents, and often alkalies, may be employed as adjuvants to the opium. The opiate and alkali may be employed in the following combination:

    1082. R. Tincturæ opii,   gtt. xij.
              Misturæ cretæ,   f. ℥jss.   M.
    One teaspoonful every two or three hours to an infant one year old.

To this mixture an astringent may be added, as tincture of catechu or kino. It should be borne in mind, however, that astringents are less tolerated by an irritable stomach than opium or chalk. When they are vomited, therefore, they should be discontinued, even in cases in which they would doubtless be serviceable if the stomach were retentive.

By means of the opiate and astringents, if they be retained, the passages are rendered, in a few hours, less frequent, and the stools more consistent.

In cases in which calomel is employed our author does not recommend its use in larger doses than one-fourth of a grain, morning and evening (together with the astringent and opiate), to a child of one year.

Dr. S. also advises small pieces of ice in the mouth at the beginning of the attack, to combat the irritability of the stomach, and the application of mustard to the epigastrium.

In most cases Bourbon whisky or brandy, the best of the alcoholic stimulants, are required. They should be used from an early period of the disease, both for the purpose of sustaintaining the vital powers and of diminishing the gastric irritability.

The diet should be simple, but nutritious, and taken often, but little at a time. If the child be at the breast, it should be confined to the mother's milk. If it be weaned, cold barley or rice water, with whisky or brandy, should be given in the commencement of the attack ; afterward, milk or broth may be employed in addition.

ALFRED VOGEL, M. D., PROFESSOR OF CLINICAL MEDICINE
IN THE UNIVERSITY OF DORPAT, RUSSIA.

Our author states that in general the rule holds good that *no child with intestinal catarrh tolerates cow's milk*, whether pure, or mixed with tea, or boiled into a broth with meat or bread, and that the diarrhœa will only exceptionally be arrested if a milk-diet is persevered in. Total abstinence from cow's milk is the first essential to successful treatment. As soon as the liquid stools appear, the patients should only be allowed demulcent drinks. In the place of milk the children may be allowed for their meals a thin mucilaginous beefbroth, with rice, barley, or groats, slightly sweetened with sugar; it should, however, be deprived of fat and without salt.

When the appetite improves a few teaspoonfuls of triturated wheat bread may be boiled in the beef broth. After the stools have been normal for at least two days, a trial may be made with one milk-pap each day, then with two, and finally three a day.

The *penciling of the mouth with laudanum,* and the use of *opiate clysters* stand at the head of all therapeutic measures. But occasionally, in the profuse diarrhœa of summer, opium proves inefficacious ; then order small doses of calomel, gr. ¼, three or four times daily, or,

1083. ℞. Argenti nitratis,   gr. ss.
Aquæ destillatæ,   f. ℥iij.   M.
A teaspoonful three or four times a day. A drop of lauda-
num may be added to each dose.

Vegetable remedies containing tannic acid, such as calumba, rhatany, pure tannic acid itself, and astringents in general, are with difficulty administered to small children, unless mixed with large quantities of syrup, and, on that account, should be seldom resorted to. In older children they may be oftener employed.

1084. ℞. Aluminis,   gr. vj.
Syrupi acaciæ,   f. ℥iij.   M.
A teaspoonful ter die.

This will sometimes check the diarrhœa which has been uninfluenced by any of the above remedies.

Our author, if compelled to choose between the two, would prefer the dietetic treatment alone to that by medicine alone. He has often convinced himself of the utter inefficacy of all therapeutic remedies in the treatment of this disease when the child is sustained on milk diet.

The best prophylaxis consists in rendering the cow's milk given the child alkaline by the addition of the following soda solution to each meal :

1085. R. Sodæ carbonatis,   ʒj.
           Aquæ,   f. ʒvj.   M.

In summer the entire quantity of milk to be consumed in the twenty-four hours should be rendered alkaline immediately upon its arrival at the house, by adding a tablespoonful of this solution to every five ounce of milk.

If this direction be followed it will become speedily evident that intestinal catarrhs may often be avoided.

PERTUSSIS.

JOHN L. ATLEE, M. D., LANCASTER, PENNA.

1086. R. Acidi hydrocyanici dil.,   ♏.j.
           Syrupi simplicis,   f.ʒj.   M.

A teaspoonful for a child six months of age, to be given at first morning and evening. If no unpleasant symptoms, dizziness, or sickness result, in forty-eight hours, the dose is to be given ter die.

One drop of the acid should be added to the recipe for each year of the child's age above one year. Our author has never given this remedy more frequently than four times a day.

DR. BARTLETT.

Our author records in the *Amer. Practitioner* for Feb., 1870, his experience with the *ioduret of silver* in whooping cough. He has used this remedy for twenty-five years, and has found that it decidedly controls the frequency and violence of the paroxysms within a week, and often within a few days, after commencing its use. The formula he employs is as follows:

1087. R. Argenti iodureti,   gr.x.
           Sacchari albi,   gr.lxx.
           Pulv. gummi tragacan.,   gr.viij-x.   M.

Rub well together, moisten with a few drops of water, make pill mass, and divide into eighty pills. Each pill will represent ¼ of a grain of the salt. One to be taken three to five times a day by a child two or three years old.

Our author generally gives them immediately before or after meals, and if the paroxysmal cough be severe, one also half way between meals. For children of from six to ten years of age two pills are to be given at a dose, and to be continued in this manner until the cough has disappeared. The remedy requires no watching, is readily taken by the youngest children, and does not interfere with the digestive functions. It is safe, pleasant, and effective.

J. HUGHES BENNETT, M. D., F. R. S. E., PROF. IN THE UNIVERSITY OF EDINBURGH.

Our author considers that whooping-cough is one of those disorders that has a natural course which is little affected by remedial measures. The efforts of the physician should be directed toward keeping the surface warm, preventing exposure to cold winds and alternations of temperature, and supporting the strength by good diet and a little wine. When the disorder becomes chronic, change of air is of undoubted service in removing the disease.

DR. GOLDING BIRD, LONDON.

1088. R. Aluminis,   gr.xxv.
          Extracti conii,   gr.xij.
          Syrupi rhœados,   f.ʒij.
          Aquæ anethi,   f ʒiij.   M.
A medium-sized spoonful every three hours in the second or
    nervous period of the disease, after the subsidence of
    inflammatory symptoms, and when the patient is harrassed
    and exhausted by the attempts to get rid of the copious
    bronchial secretion.

Under these circumstances, our author considers alum, which he administers according to the above formula—the most satisfactory of all remedies, affording the speediest and most marked relief.

BEDFORD BROWN, M. D., OF ALEXANDRIA, VA.

Our author recommends the *oil of turpentine* in the treatment of certain complications of whooping-cough. The remedial influence of this agent is that of a soothing, but active expectorant, reducing inflammatory action, tranquilizing irritation, and diminishing copious and exhausting discharges. The morbid complications of pertussis to which the turpentine treatment is more particularly adapted, are infantile, remittent, or gastric fever, a peculiar irritative form of continued fever, bronchitis, hemoptysis, convulsions, dysentery and enteritis. The doses our author employs, and the manner and frequency of administration by which he has obtained his result, he does not record.

JOHN COOPER, M. D., PHILADELPHIA.

Dr. C. has used, with excellent success, the *tonka bean* in pertussis. He was led to its employment by the fact that it contains a large per centage of *coumarin*, the active principle of the clover tops—*trifolium mellilotus*—recommended for the affection. He gives the *fluid extract* in from five to eight drop doses every three hours, for a child of five years. It affords marked relief to the paroxysm of coughing.

M. LE. DR DAVREUX, OF LIÉGE.

The following formula is recommended as a prophylactic and abortive remedy in whooping cough :

1089. ℞. Extracti aconiti,   gr j.
           Aquæ laurocerasi,   f.ℨj.
           Syrupi ipecacuanhæ,   f.ℨj.
           Mucilaginis acaciæ,   f ℥vj.   M.
Dose—f.ℨj-ij every hour for a child, and f.℥ss for an adult.

J. LUDLOW, M. D., CINCINNATI, OHIO.

Dr. L. has found *chestnut leaves* (castanea vesca), a valuable remedy for this disease. The spasm is relieved in from five to ten days, and in about two weeks the little sufferer ceases to whoop, and goes on to a speedy recovery. He employs an infusion made as follows:

1090. R.   Castanæ vescæ,   ℥ss.
            Aquæ bullientis,   Oj.   M.

Add to this a pint of cold water. Sweeten with white sugar to make palateable and administer cold. As much should be given during the day and evening as the patient can be induced to take.

By offering it as a drink in place of cold water the child soon learns to like it, and there is no trouble in getting a sufficient quantity taken to produce the desired result.

DR. MACKELCAN, CANADA.

1091. R.   Potassii sulphureti,   gr.xxiv.
            Syrupi,   f.℥j.
            Aquæ destillatæ,   f.℥iij.   M.
       A tablespoonful ter die

The dose should be increased one grain for each year up to four years of age, and after that half a grain additional for each year, the doses being diluted in proportion to the quantity of the salt.

The beneficial effects of the medicine are not perceived for five days, when the intervals between the paroxysms of cough become longer, and after that their violence diminishes froms day to day, until at the end of ten or fourteen days it is seldom necessary to pursue the treatment further.

As the drug easily spoils by keeping, it is important to have it fresh. If it dissolves perfectly in the syrup and water, and the mixture is of a greenish color, it may be relied on; but if there is any sediment, it has been decomposed by exposure to air and becomes a sulphate.

DRS. MEIGS AND PEPPER, PHILADELPHIA.

1192. ℞. Aluminis, ℈ijss.
    Syrupi zingiberis,
    Syrupi acaciæ,
    Aquæ, aa f.℥j. M.

A tablespoonful ter die every five or six hours.

This recipe, when prepared with good syrups, tastes very much like lemonade, rendering it acceptable to children.

Our authors more generally employ alum in combination with belladonna. They have obtained better results from the following formula than any other ever employed:

1093. ℞. Extracti belladonnæ, gr.j.
    Aluminis, ℥ss.
    Syrupi zingiberis,
    Syrupi acaciæ,
    Aquæ, aa. f.℥j. M.

A teaspoonful morning, noon and night, also once in the night if the cough be troublesome.

1094. ℞. Potassæ carbonatis, ℈j.
    Cocci, ℈ss.
    Sacchari albi, ℥j.
    Aquæ, f.℥iv. M.

Dessertspoonful ter die to a child a year old.

This mixture has long enjoyed a high reputation in this country and abroad. Our authors believing its efficacy to be due to the carbonate of potash, ordinarily omit the cochineal. This recipe, together with the alum and belladonna mixture given above, are the most useful agents we have to keep down the violence of the disease.

GEO. W. SMITH, M. D., ALABAMA.

1095. ℞. Syrupi scillæ compositi, f.℥j.
    Antimonii et potass. tart., gr.j.
    Tincturæ tolutanus,
    Tincturæ cocci, aa. f.℥ij.
    Mellis, f.℥ij. M.

A teaspoonful every two hours for a child three years of age; or a smaller dose at shorter intervals.

Our author has tested this remedy in many cases characterized by severe paroxysms, and with the most happy results.

J. LEWIS SMITH, M. D., PROF. BELLEVUE HOSPITAL COLLEGE, NEW YORK.

In the catarrhal stage the treatment of pertussis should not differ from idiopathic catarrh. Mild counter irritation to the chest, and, if there be accelerated breathing, the oil silk jacket may be applied, while demulcent, laxative, and gentle expectorant mixtures are employed internally. Care should be taken not to reduce the strength nor impair the general health.

In the second stage, that of convulsive cough, therapeutic measures are most beneficial. The three medicines most in favor with the profession at the present day are *hydrocyanic acid, belladonna* and *bromide of ammonium.* In the opinion of our author, the treatment by belladonna is the most successful.

The first dose of belladonna should be smaller than that which will prove remedial. The child requires a larger dose of belladonna, however, than the adult in order to produce the same effect.

Our author recommends the following pills, directed by Trousseau :

> 1096.   R.   Extracti belladonnæ,
>   Pulv. belladonnæ folii,   aa. gr.ij.   M.
>   For xx pills.

One of the pills, (containing a tenth of a grain each of the extract and of the leaves of belladonna) should be taken in the morning when the stomach is empty, and a second on the following morning. For children over four years the strength

of the prescription is to be doubled (*i. e.*, gr. 1-5 of each ingre-
dient in every pill). If the number of paroxysms be diminished or the cough rendered less severe, the same dose is to
be administered each day. If, however, there be no improvement, two pills are to be administered on the second morning,
three on the next, and so on until an appreciable effect is
produced. Trousseau considered it important to give at one
dose whatever belladonna is administered during the day;
divided doses are less effectual.

The dose which produces amelioration of the symptoms is be
repeated daily during the succeeding six or eight days. Then,
if the improvement continue, the dose is to be diminished one
pill per day back to the first dose, but if the cough increase,
it is to be again increased. After the entire cessation of the
spasmodic cough the administration of the remedy is to be
kept up for six or eight days. Instead of belladonna, atropia
may be used in the same manner in solution.

> 1097. R. Atropiæ sulphatis,   gr.1-10.
>   Aquæ,   f.℥ij.   M.
>
> A teaspoonful for young children, employed as directed for
>   the belladonna pills. Older children require a propor-
>   tionately larger dose.

Our author commonly employs the *extract of belladonna* in
one-grain pills. For an infant one year old, one pill is dissolved in eight teaspoonfuls of water; three years, in four
teaspoonfuls. A teaspoonful to be given once, or, if there be
no appreciable effect, three or four times daily. If there be no
modification of symptoms an additional half spoonful should
be given on the third day. Afterward a still further increase
will probably be required.

24

FORMULÆ AND DOSES OF MEDICINES FOR HYPODERMIC MEDICATION.

ARSENIC.

DR. C. B. RADCLIFFE.

1098. R. Liq. potassæ arsenitis,
Aquæ destillatæ,   aa. ♏.iij.   M.
For one injection, gradually increased to ♏.xiv of Fowler's solution.

PROF. ROBERTS BARTHOLOW suggests the *liquor sodæ arsenitis* (in doses of ♏.v., x. or even xv., on every other day) as less irritating than Fowler's solution.

Therapeutics: Dr. RADCLIFFE has used arsenic hypodermically, with benefit, in cases of chorea, neuralgia, epilepsy and other nervous affections.

ATROPIA.

FORMULA FOR THE SOLUTION.

PROF. ROBERTS BARTHOLOW, CINCINNATI, OHIO.

1099. R. Atropiæ sulphatis,   gr.ij.
Aquæ destillatæ,   f.℥j.   M.
5 minims = gr. 1·48.

With this formula the dose can be better regulated than with stronger solutions.*

REMARKS ON THE HYPODERMIC USE OF ATROPIA.

Dose.—Lorent begins with gr. 1-30, and goes up to gr. 1-20 ; Sudekum and Behrer, gr. 1-60 ; Hunter, gr. 1-25 ; Scholz and Oppolzer, gr. 1-20 ; Græfe and Depuit, gr. 1-12 ; Nudirfer gr. 1-10 ; Courty,' gr. 1-6 ; Bell, gr. 1-4 ; Trousseau, gr. 1-12

---

*Hypodermic Medication. Lippincott, Philada , 1865, pp. 75, 89.

to 1-6; Ruppaner, gr. 1-60 to 1-30. Dr. BARTHOLOW says that 5 ℳ. of (F. 1099), or 1-48 of a grain, is the largest amount desirable in most cases, and that it will be rarely necessary to inject more than gr. 1-24 at one time.

M. BEHIER.

1100. ℞. Atropiæ valerianatis, gr.v.
Aquæ destillatæ, f.ℨj. M.

Our author has injected this solution (gtt. v. every two hours) along the nape of the neck in *tetanus*.*

ATROPIA AND MORPHIA.

PROF. ROBERTS BARTHOLOW.

1101. ℞. Morphiæ sulphatis, gr.xvj.
Atropiæ sulphatis, gr.j.
Aquæ destillatæ, f.ℨj. M.
Filter.

5 minims = gr. 1-6 of morphia and gr. 1-96 of atropia. Or, combine f.ℨj. of F. 1099 with f.ℨiv. of F. 1101, making a solution of which 5 minims = gr. ¼ of morphia, and gr. 1-95 of atropia.

See also F. 64 and 109.

Therapeutics: Used in insomnia, (in the proportion of gr. 1-120 to 1-96 of atropia to gr. ¼ to ½ of morphia†); neuralgia (F. 109); epilepsy (F. 64); asthma; angina pectoris, spermatorrhœa (atropia in excess); pelvic and uterine pain; rheumatic arthritis; muscular and acute rheumatism (in all such cases atropia in excess).

---

*The *Medical Record*, N. Y., 1869. p. 173.
†Bartholow on Hypodermic Medication, p. 95

CAFFEIN.

DR. ALBERT EULENBURG, BERLIN.

1102.   R.   Caffeini puri,   gr.vj·
              Alcoholis,
              Aquæ destillatæ,   aa. f.ʒj·   M.

        20 minims = gr. j.

DR. E. LORENT, BREMEN.

1103.   R.   Caffeini citratis,   gr j.
              Glycerinæ,   gtt·xxiv.   M.
For one injection.

REMARKS ON THE HYPODERMIC USE OF CAFFEIN.

Dose—gr.j.

Therapeutics: In *neuralgia, hysterical headache,* and *opium poisoning.* Prof. Bartholow suggests that as there is no incompatibility, caffein and atropia be used at the same time hypodermically in cases of opium narcosis.* Dr. Eulenburg states that caffein, when injected in doses from one-fifth to two-thirds of a grain, relieves *occipital neuralgia* and *hysterical headaches* generally.

CONIA.

FORMULA FOR THE SOLUTION.

DR. A. ERLENMEYER.

1104.   R.   Coniæ,   gr.ij·
              Alcoholis,   f.ʒij.   M.
Dissolve and add

              Aquæ destillatæ,   f.ʒij.   M.
              gtt.j = gr. 1-120.

---

*Manual of Hypodermic Medication, pp. 125, 126.

DR. ALBERT EULENBURG, BERLIN.

1105. ℞. Coniæ,   gr.ss.
           Alcoholis,   f̃ʒss.   M.
Dissolve and add
           Aquæ destillatæ,   f.ʒiss.   M.
           5 minims = gr. 1-48.

As these solutions quickly decompose, they should be freshly made for use.

REMARKS ON THE HYPODERMIC USE OF CONIA.

Dose: This ranges from gr. 1-120 to 1-60.

Therapeutics: This drug has been employed in the treatment of *tetanus, asthma, emphysema, angina pectoris,* etc.

DATURIA.

1106. ℞. Daturiæ,   gr.j.
           Aquæ destillatæ,   f.ʒij.   M.
           gtt.iv. = gr. 1-30.

DIGITALIN.

DR. ALBERT EULENBURG, BERLIN.

1107. ℞. Digitalin,   gr.ss.
           Alcoholis,
           Aquæ,   aa. f.ʒj.   M.
           gtt. iv. = gr. 1-60.

DR. ULLERSPERGER, OF MUNICH.

1108. ℞. Digitalin,   gr.j.
           Glycerinæ,
           Aquæ destillatæ,   aa. f.ʒij.   M.
           gtt. iv. = gr. 1-60.

The *doses* employed have been as follows:

Eulenburg, gr. 1-100—1-60; Franque, gr. 1-20—1-10; Lorent, gr. 1-10—1-5; Pletzer, gr. 1-20.

ERGOT.

DR. ALBERT EULENBURG, BERLIN.

1109. R. Ergotini,   gr.ij.
       Alcoholis,
       Glycerinæ,   aa. f.ʒss.   M.
       5 minims = gr. 1-6.

FREDERICK D. LENTE, M. D., COLD SPRING, NEW YORK.

1110. R. Extracti ergotæ fluidi,   ℧.xv.
For one dose.

Our author reports a case of post-partum hemorrhage treated in this way.*

REMARKS ON THE HYPODERMIC USE OF ERGOT.

Dose: About gr. 1-6 of ergotin; from ℧.x–xv. or more of the fluid extract of ergot (U. S. P.)

Therapeutics: Used in *post-partum hemorrhage, epistaxis,* etc., and in *internal aneurism.* (See p. 395.)

ACIDUM HYDROCYANICUM DILUTUM.

PROF. ROBERTS BARTHOLOW.

1111. R. Acidi hydrocyanici dil.,   ℧.ij–iv.
For one injection.

For ordinary purposes the smaller dose should be preferred. It may be frequently repeated, as its influence is soon dissipated.

REMARKS ON THE HYPODERMIC USE OF HYDROCYANIC ACID.

Dose: ℧.ij–iv of the officinal dilute acid, (U. S. P.)

Therapeutics: This remedy is useful in *functional nausea* and *vomiting,* in *gastralgia* and in *mental disorders.*†

New York Medical Record, 1869, p. 411.
†Bartholow on Hypodermic Medication, p. 117.

MERCURY.

PROF. ROBERTS BARTHOLOW.

1112. ℞. Hydrarg. chloridi corros., gr.j.
Aquæ destillatæ, f.℥j. M.
10 minims = gr. 1-48.*

For M. BOUILHON's formula for the double iodide of sodium and mercury, see F. 992.

For remarks concerning the hypodermic use of mercury in syphilis, see p. 315.

MORPHIA.

MORPHIA ACETAS—FORMULÆ FOR THE SOLUTION.

MIDDLESEX HOSPITAL, LONDON.

1113. ℞. Morphiæ acetatis, gr.x.
Acidi acetici, ℳj.
Aquæ, q. s. ad f.℥j.
Liquoris potassæ, ℳj. M.
1 minim = gr. 1-6.†

PLETZER uses the following solution :

1114. ℞. Morphiæ acetatis, gr.iij.
Aquæ destillatæ, f.℥j. M.
5 minims = gr. ⅓.‡

DR. E. LORENT, BREMEN.

SOLUTION No. 1.

1115 ℞. Morphiæ acetatis, Əj.
Aquæ destillatæ, f.℥j. M.
6 minims = gr. ⅓.

SOLUTION No. 2.

1116. ℞. Morphiæ acetatis, Əj.
Aquæ destillatæ, f.℥ij M.
6 minims = gr.j.

---

*Manual of Hypodermic Medication, p. 133.
†Squire's Pharm. of the London Hospitals, London, 1869, p. 47.
‡Ullersperger's Prize Essay, Transactions of Pennsylvania State Med. Society, 4th series, part iii. p. 457.

SOLUTION No. 3.

1117. ℞. Morphiæ acetatis,   gr.xxx.
    Aquæ destillatæ,   f.ℨij.   M.
   6 minim = gr.iss.

MORPHIÆ MURIAS—FORMULA FOR THE SOLUTION.

DR. ALBERT EULENBURG.

1118. ℞. Morphiæ muriatis,   gr.iv.
    Acidi muriatici,   gtt.iv.
    Aquæ destillatæ,   f.ℨj.   M.
   3 minims = gr. 1–5.*

The objection to this solution is its *acidity*, which often provokes pain and local irritation.

MORPHIÆ SULPHAS—FORMULA FOR THE SOLUTION.

PROF. ROBERTS BARTHOLOW.

1119. ℞. Morphiæ sulphatis,   gr.xvj.
    Aquæ destillatæ,   f.ℨj.   M.
   Dissolve and filter.
   5 minims = gr. 1–6.

The advantage of this solution is that it contains no acid and pure water causes very little irritation.†

REMARKS ON THE HYPODERMIC USE OF MORPHIA.

Dose: The dose is variously given by different authorities. The age, sex, constitution, temperament and the nature of the disease, all, of course, influence the amount of each injection, as well as the frequency of administration. Drs. E. LORENT and SCHOLTZ have used as high as gr. jss. at a single injection.

Dr. BARTHOLOW varies the dose from gr. 1-12 to gr. ½. He says that "*in commencing it should not exceed one-third of that ordinarily administered internally.*" He regards the large doses (gr. ½, ⅓ and j.) as unsafe for the first trial, unless the conditions requiring the injection be exceptional. Dr. RUP-

*Bartholow on Hypodermic Medication, p 35.
†Hypodermic Medication, p. 36.

PANER places the minimum dose at gr. ⅓, the maximum at gr. ¾. Dr. EDWARD JOHN TILT says that the initial dose for a woman should never exceed gr. 1-6. Dr. CHAS. HUNTER, of London, gives the rule never to use in the first injection more than one-half the stomachic dose for males, and not more than a third for females.

Therapeutics: Morphia is used hypodermically in neuralgia (F. 109). Delirium tremens, hysteria, epilepsy, insomnia, chorea, tetanus, hydrophobia, asthma, catarrh, emphysema, pleurisy, dyspepsia, cholera, colic, vomiting of pregnancy, urinary affections, and as an antidote to the toxic effects of atropia, strychnia and digitaline. Dr. Bartholow considers it inferior in strychnia poisoning to the calabar bean (see F. 1121); and in poisoning by digitaline as less efficacious than atropia.*

Cautions: Dr. NUSSBAUM has forcibly indicated the danger that may arise from the penetration of a superficial vein by the point of the syringe. His experience may serve to put physicians on their guard against a fearful danger. He observes: "During the last two months I have undergone a frightful experience twice in my own person, and three times in the case of my patients. The point of the syringe entered a subcutaneous vein, and the morphia was thus injected directly into the blood, instead of into the subcutaneous tissue. On the first occasion I injected two grains of acetate of morphia, (†) dissolved in fifteen minims of water, into one of my subcutaneous abdominal veins, and felt as if I should die in a few minutes. In a couple of seconds there was a pricking and burning sensation over my whole body, a strongly acid taste in my mouth, my whole face was nearly as red as the

---

*Manual of Hypodermic Medication, pp. 37, 38, and 70.
†This would have proved a highly dangerous dose to most persons if simply injected into the subcutaneous tissue.

normal color of the lips, and in about four seconds after the
injection there was a ringing sound in the ears, while scintil-
lations flashed before the eyes, and there was intense pain in
the integuments of the head. But the most terrible of all the
phenomena was the extremely powerful and rapid action of
the heart. Out of more than 25,000 patients I have never
felt such a pulse. Its beats ranged from 160 to 180 in the
minute, while the carotids had no time to discharge their contents, and felt like thick tremulous iron cords on either side of
the neck. The action of the heart and arterial pulsations
was so strong that I felt as if the walls of the chest or the
diaphragm must give way, and that my eyeballs must burst.
This fearful state, in which the respiration was considerably
impeded, lasted on the first occasion about eight minutes·
The suffusion of the face was followed by a deadly pallor,
which lasted for an hour, while the acute pain in the head
subsided in fifteen minutes. The mind was in no degree affected, and with an effort I could stand and speak. Cold applied in the form of washing, affusion, etc., was very agreeable and beneficial. In the course of two hours the whole of
the symptoms disappeared. In my other personal misadven-
ture the symptoms were far less severe in consequence of the
injected dose being much smaller. Taught by experience, I
have since then always injected very slowly, and as the phenomena come with such lightning-like rapidity, I thus secure
time, if necessary, to reverse the pumping action of the syringe
and to recover a part of the injected fluid mixed with blood.
I have on several occasions seen the happy results of this
manipulation. The three of my patients in whom a vein was
entered were in even a more critical state than I personally
was. There was a partial loss of consciousness, and there
were convulsions, but no persistent consequences ensued."

Dr. EULENBERG, of Berlin, with an experience of many thousand cases of injection, has never met with this accident, but does not, on that account, call into question the accuracy of NUSSBAUM's statements. Inflammation of the punctured spot has been noticed by Dr. E. on only three occasions, in all of which it was clearly due to the irritant nature of the injected fluid.*

NICOTIA.

DR. A. ERLENMEYER.

1120. R· Nicotiæ,   gr.ss.
         Aquæ destillatæ,   f.ʒij.   M.
     4 minims = gr. 1–60 *

REMARKS ON THE HYPODERMIC USE OF NICOTIA.

Dose—gr. 1–60.

Therapeutics: Prof. Houghton, of Dublin, has employed this agent with success in cases of traumatic tetanus, of which about one-half the cases treated recover. This result is better than that obtained from any other drug excepting the calabar bean (see p. 40). Nicotia is a physiological antagonist to strychnia.†

PHYSOSTIGMA.

PROF. ROBERTS BARTHOLOW.

1121. R· Extracti physostigmæ,   gr.ij.
         Aquæ destillatæ,   f.ʒj.   M.
     Filter.
     10 minims = gr.⅓·

(For the method of making the extractum physostigmæ, see F. 104.)

* *Medical Times and Gazette*, Oct. 39th, 1869, p. 526
†Ullérsperger's prize essay, op. cit.

This solution must be prepared when wanted, as it soon becomes unfit for use. Its acidity should be neutralized by carbonate of soda.

REMARKS ON THE HYPODERMIC USE OF CALABAR BEAN.

Dose: gr.⅓ of the extract to begin with.

Therapeautics : Tetanus and chorea have both been treated with success by this remedy. In the first named affection it probably stands at the head of all known remedial agents. It is also employed in *strychnia poisoning.*

Dr. ALOIS MONTI, of the St. Ann's Child's Hospital, reports three cases out of five *of trismus neonatorum* cured by this remedy. He prefers subcutaneous injection, as he thinks the internal use uncertain. He repeats these injections every ten or fifteen minutes until the spasms cease ; then intermits them, even for several hours, until the cramps return again. For new-born children he uses one-tenth grain of the extract per dose, and goes up to one-third, one-half, or a whole grain a day. Older children can commence with one-third grain per dose. For internal use from one to four grains a day may be given.

The *antidote* to physostigma is strychnia, which is its physiological antagonist.

QUINIA.

FORMULÆ FOR THE SOLUTION.

PROF. ROBERTS BARTHOLOW, CINCINNATI, OHIO.

1122.   R.   Quiniæ sulphatis,   ʒj.
           Acidi sulphurici diluti,   ℳ.xl.
           Aquæ destillatæ,   f.℥j.   M.
    Dose 15 to 30 minims.
      Carefully filter.

---

*Bartholow on Hypodermic Medication, p. 115.

Inject where the areolar tissue is abundant.

1123. R. Quiniæ, gr.viij.
.Etheris, f.ℨj.* M.

DR. DESVIGNES.

1124. R. Quiniæ, gr.iss.
Acidi nitrici diluti. ♏.j.
Aquæ destillatæ, ♏.xv. M.
For one injection.

Our author has treated several hundred cases of intermittent fever with this injection. The patients were railroad laborers working in the Tuscan salt marshes.†

DR. ADDINELL HEWSON, PHILADELPHIA.

1125. R. Quiniæ sulphatis, gr.iij.
Aquæ destillatæ, f.ℨj.
Acidi sulphurici diluti, q. s. M.
To make a neutral solutior.
10.gtt. = gr.½.

Two cautions should be noted, i. e., to have this solution *neutral,* and to have it *freshly made.* If these be observed, there is no danger of the formation of an abscess from its use. Ten drops may be thrown under the skin three times a day. Dr. HEWSON has employed this injection as a tonic in bad cases of typhoid fever with signal success. He has also found it valuable in breaking up obstinate intermittents, when the ordinary treatment by the mouth failed.

REMARKS ON HYPODERMIC USE OF QUINIA.

In the intermittent fever arising from malaria, the hypodermic injection of quinine is a preëminently successful mode of treatment. Dr. SCHACHANA, of Smyrna, states that a single application suffices to effect a cure, good diet and chalybeates being also prescribed. Out of 150 cases there was

*Manual of Hypodermic Medication, p. 120.
†*Medical Times and Gazette,* Oct. 30, 1860, p. 529.

only one relapse. GUALLA, of Breschia, similarly treated forty-nine cases without a single failure. DESVIGNES treated several hundred cases occuring in navvies engaged on railway work in the Tuscan salt marshes, and met with uniform success. Dr. EULENBURG injected quinine in eleven cases of intermittent fever, and confirms the view propounded by previous observers, that this medicine, when injected in doses of one and a half or two grains before or during the cold stage, has the power of cutting short the attack. Five of his eleven patients complained of a sharp burning pain while the fluid was being injected, and for some minutes subsequently.

In his remarks upon "intermittent and remittent fever, independent of malaria," Dr. EULENBURG states that he convinced himself by many accurate observations that "by the subcutaneous injection of small quantities of quinine, we are able in a great number of febrile states of a remittent or intermittent type to produce a temporary, and frequently a considerable, diminution of the febrile temperature of the body." This fact, which he clearly proves by numerous cases, obviously has an important bearing upon the treatment of various forms of disease. In cases of typical neuralgia (sciatica and tic) this remedy has been highly serviceable.

Dr. ADDINELL HEWSON, of Philadelphia has found it valuable in the treatment of typhoid fever and as a tonic. (F. 1125.)

STRYCHNIA.
FORMULÆ FOR THE SOLUTION.
PROF. ROBERTS BARTHOLOW, CINCINNATI, OHIO.

1126. R. Strychniæ sulphatis, gr.ij.
Aquæ destillatæ, f.ℨj. M
5 minims = 1–48.*

*Manual of Hypodermic Medication, p. 99.

DR. E. A. ERLENMEYER.

1127. ℞. Atropiæ sulphatis, gr.j.
Aquæ destillatæ, f.ʒij. M.
5 minims = gr. 1-24.

DR. ALBERT EULENBURG, BERLIN.

1128. ℞. Strychniæ sulphatis, gr.ij.
Aquæ, f.ʒij. M.
1 minim = gr. 1-60.

DR. WALDENBURG.

1129. ℞. Strychinæ sulphatis, gr.ij
Glycerinæ, f.ʒss.
Aquæ destillatæ, f.ʒiss. M.
1 minim = gr.1-60.*

All of these solutions become unfit for use if long kept on hand.

REMARKS ON THE HYPODERMIC USE OF STRYCHNIA.

Dose: This alkaloid has been employed in various doses
thus : Neudorfer, gr.1-40 ; Echeverria, of New York, gr.1-60
to 1-30 ; Charles Hunter, gr.1-90 to 1-24 ; Bartholow, gr.1-48
to 1-24 ; Waldenburg, and Delbeau, gr.1-10 ; Eulenburg, and
Bois, gr.1-8 ; Courty, gr.1-6 ; Ruppaner, gr.1-24 to 1-6 ;
Lorent, gr.1-25 to 1-10.

ANTIDOTE FOR STRYCHNIA POISONING.

The *calabar bean* (F. 1122) is a complete antagonist to the
toxic effects of strychnia. In its absence, the inhalation of
ether, successfully employed by Dr. Echeverria may be resorted to. Dr. Eulenburg mentions a case that occurred at
Königsburg, in which a young man who had taken one gram.
and a-half of strychnia was apparently saved by the hypodermic application of *woorara.**

Therapeutics: The subcutaneous injection of strychnia is
principally used in cases of paralysis and neuralgia. Anstie

*Ullérsperger's Prize Essay, Transactions of the Medical Society of the State
of Pennsylvania, 1869. p. 468.

commends it as the remedy *par excellence* in *gastralgia*, injected in doses of gr.1-120 to 1-60. Dr. Eulenburg has found it (in doses from 2-25 to 4-26 of a grain of the sulphate) highly valuable in cases of facial paralysis, paralysis of the vocal cords, paralysis of the bladder, prolapsus, spinal paraplegia, spasmodic muscular contractions, amaurosis and sciatica.

WOORARA.

FORMULÆ FOR THE SOLUTION.

DR. SCHUH, VIENNA.

1130. ℞. Woorarae,   gr.j.
Alcoholis,   gtt.clx.   M.
8 minims = gr. 1-20.

REMARKS ON THE HYPODERMIC USE OF WOORARA.

Dose: This may be said to vary between gr. 1-60 and gr. 1-20. SPENCER WELLS has injected as much as gr. 1-12 at one time. Gherini had gr.ij. dissolved in f.ʒij. aquæ dest., and injected the solution in twenty-four hours.

Therapeutics : Tetanus is the disease for which woorara has been chiefly used subcutaneously. It has also been administered in *epilepsy*.

FORMULÆ AND DOSES OF MEDICINE FOR INHALATION.

The doses are calculated for an ordinary steam atomizer.

1131. ℞. Acidi carbolici fluidi,   gtt.iij–x to aq. f.ʒj.
In phthisis.

1132. ℞. Acidi tannici,   gr.j-xx to aquæ f.ʒj.
In chronic catarrhal affections, œdema of glottis, and laryngeal ulcerations. In ordinary laryngitis and in bronchitis,.

begin with small doses and discontinue if much heat and dryness be produced (Da Costa).

1133. ℞. Aluminis, gr.v–xxx to aq. f.ʒj.

Particularly useful in cases of excessive secretion from bronchia (Da Costa). In large doses employed in pulmonary hemorrhage. More sedative and better suited to irritable conditions than tannin.

1134. ℞. Ammoniæ muriatis, gr.ij–ʒij to aq. f.ʒj.

To promote expectoration in acute and chronic laryngeal and bronchial catarrh, and in capillary bronchitis. Siegle says the dose best borne is not above gr.x. to f.ʒj.

1135. ℞. Aquæ destillatæ, f.ʒj.

Warm in inflammatory and spasmodic affections; cold in hemorrhage.

1136. ℞. Aquæ amygdalæ amaræ, f.ʒj.

A sedative in painful affections of upper air passages and paroxysmal cough.

1137. ℞. Aquæ assafœtidæ, f.ʒj.

Used in asthma with emphysema.

1138. ℞. Aquæ calcis, f.ʒj.

In diphtheria and membranous croup.

1139. ℞. Aquæ picis liquidæ, f.ʒj–ij to aquæ f.ʒj.

In offensive bronchial secretions; in gangrene of the lungs; and in tuberculosis.

1140. ℞. Argenti nitratis, gr.j–x to aquæ f.ʒj.

In ulcerations and in follicular pharyngitis a face shield always to be worn. The largest dose only in cases of ulceration.

1141. ℞. Atropiæ sulphatis, gr.1-40 to aq. f.ʒj.

A dangerous inhalation.

25

1142. ℞. Cadinii olei., gtt.j–ij to aq. f.℥j.

In the chronic catarrh of emphysema.

1143. ℞. Cannabis ind. ext., gr.¼–j to aquæ f.℥j.

In spasmodic and irritative coughs ; phthisis.

1144. ℞. Cannabis ind. tinct., ℳ.v–x to aq. f.℥j.

Uses : same as of extract.

1145. ℞. Conii extracti, gr.j–vj to aq. f.℥j.

In irritative coughs and in asthma.

1146. ℞. Conii ext. fluidi, ℳ.iij-viij to aq. f.℥j.

Used for the same purposes as above.

1147. ℞. Cupri sulphatis, gr.j–xx. to aq. f.℥j.

In chronic inflammations and ulcerations.

1148. ℞. Ferri lactatis, gr.j–ij to aq. f.℥j.

In anæmia (see F. 544.)

1149. ℞. Ferri chloridi, gr.½–ij to aq. f.℥j.

In the earlier stages of phthisis and in hysterical aphonia. To be used stronger in chronic pharyngitis and laryngitis. In pulmonary hemorrhage, gr.ij–x to aquæ f.℥j ; or,

1150. ℞. Ferri sulphatis liq., ℳ.x–xl to aq. f.℥j.

In pulmonary hemorrhage.

1151. ℞. Hyosciami extracti, gr.½ to aquæ f.℥j.

In whooping cough and spasmodic coughs. The strength of this solution may be gradually increased. The *fluid extract* may be used in doses of ℳ.iij–x to f.℥j.

1152. ℞. Iodinii tincturæ, gtt.j–xx to aq. f.℥j.

In inflammatory affections of the larynx and pharynx.

1153. ℞. Iodinii liq. compositi, ℳ.ij–xv ta aq. f.℥j.

In chronic bronchitis and in phthisis.

1154. ℞. Liq. potassæ arsenitis, ℳ.j-xx to aquæ f.℥j.
Nervous asthma.

1155. ℞. Liq. sodæ chlor., f.℥ss-f.℥j to aq. f.℥j.
In phthisis and in the offensive and copious expectoration
of chronic bronchitis.

1156. ℞. Morph. acetatis, gr.1-12-⅛ to aq. f.℥j.
In irritative coughs, and for its constitutional effects.

1157. ℞. Opii extracti, gr.¼-½ to aq. f.℥j.
Used for the same purposes as F. 1156.

1158. ℞. Opii tincturæ, gtt.iij-x to aq. f.℥j.
Employed for the same affections as F. 1156.

1159. ℞. Plumbi acetatis, gr.iij-x to aq. f.℥j.
In obstinate, troublesome colds, not yielding to other medi-
cament.

1160. ℞. Potassæ carbonatis, gr.x-ℨij to aq. f.℥j
Same as ammoniæ murias. Particularly useful in follicular pharyngitis.

1161. ℞. Potassæ chloratis, gr.x-xx to aq. f.℥j.
In chronic and subacute catarrhal affections, particularly
when there is a feeling of dryness.

1162. ℞. Potassii bromidi, gr.j-x to aq. f.℥j.
In laryngeal croup.

1163. ℞. Potassii iodidi, gr.ij-xx to aq. f.℥j.
In granular inflammations. In chronic bronchitis with
emphysema.

1164. ℞. Sodii chloridi, gr.v-xx to aq. f.℥j.
In phthisis. It promotes expectoration and diminishes
sputa.

1165. ℞. Terebinth. olei rect., gtt.j-ij to aq. f.℥j.

In chronic bronchitis, with offensive secretions ; bronchor-rhœa ; gangrene of the lungs.*

THE NEW HYPNOTIC.

HYDRATE OF CHLORAL.

The profession is indebted to Dr. Oscar Liebreich, of Berlin, for a new hypnotic and anæsthetic, to which the name of hydrate of chloral has been given. Its chemical formula is $C_4 HCl_3 O_2 + HO$. The following recipe may be used for its administration internally :

> 1166. ℞ Chloral hydratis,   ℥ss.
>   Syrupi tolutanus,
>   Aquæ,   aa. f.℥iij.   M.
>   One to four tablespoonfuls for a dose, with water.

This solution has a fruity odor and sharp pungent taste. Chloral is also administered subcutaneously.

This drug acts as a hypnotic, anodyne, and antispasmodic, in cases in which other remedies have failed. The particular advantage claimed for it is, that its administration is not fol-lowed by the unpleasant after-effects which are consequent upon the use of opium and other narcotics.

Dr. Spencer Wells reports† that he has arrived at the rule that thirty grains of hydrate of chloral give as much re-lief as one grain of opium, and that the effects are more im-mediate. On the succeeding day no ill results are observed, while after opium there is invariably loss of appetite and more or less headache.

---

*Inhalations in the treatment of diseases of the Respiratory Passages. J. M. Da Costa, M. D., Philadelphia, p. 51, et seq.
    Inhalation. Its Therapeutics and Practice. J. Solis Cohen, M. D., Phila., p. 89, et seq.
    On the Inhalation of Atomized Fluids. II. Beigel, M. D., London *Lancet.*
†*Medical Times and Gazette* for September 18, 1869.

M. DEMARQUAY, after an extensive experience with this remedy, comes to the following conclusions : *

1. Chloral has a well marked hypnotic action, especially in weak and debilitated patients.

2. The duration of its action is in direct proportion to the feebleness of the patients.

3. The sleep which it brings on is generally calm, and is accompanied by restlessness only when the patients are suffering from intense pain. This causes one to employ it in diseases where it is desired especially to induce sleep and muscular relaxation.

4. This remedy may be employed in high doses, since no ill effects result from it, when administered in doses of from one to four scruples.

M. BOUCHUT,† in a report read before the Académie des Sciences, condemns, as dangerous, the use of the hydrate of chloral by hypodermic medication, and states that it is more rapidly absorbed by the rectum than by the stomach.

He pronounces it to be the most valuable sedative known in the violent pain of gout, or nephritic colic, and of dental caries ; and the quickest remedy in aggravated chorea to quiet the restlessness, which is the most serious symptom of that disease.

Prof. LANGENBECK has used it with success in cases of delirium tremens.

Dr. J. M. DA COSTA has employed it in lead colic with good results. He has found it objectionable in cases of weak heart. In such patients it produces a good deal of disturbance, and ought not to be prescribed. He thinks that the

---

*London Lancet, September 25th and October 9th, 1869.
†New York Medical Gazette for Dec. 11th, 1869.

general statements which have been published by European writers, in regard to this remedy, are about correct.

In this connection we may state that Dr. DA COSTA has ascertained that in cases in which opium could not be taken, the bromide of potassium previously given exerts a corrective influence, enabling the patients to tolerate the opium. He gives the bromide of potassium in advance so as to introduce from forty to sixty grains before ordering the opiate. In a number of instances, he has thus been able to give the latter drug where, on account of pain, it seemed absolutely indicated, but where it had habitually disagreed. The combination of the two articles answers, but not so well as the administration of the bromide of potassium separately and in advance of the opium.

INDEX OF REMEDIES.

Ammonii iodidum, F. 240, 345, 346, 1006.

Amygdalæ amaræ emulsum, F. 570, 588, 597, 670, 676.

Amygdalæ mistura, F. 240.

Amygdalæ oleum, F. 98, 628, 637, 688, 691, 718, 742, 778, 795.

Analine, F. 83.

Anisi oleum, F. 446.

Anthemis, F. 746, 751, 757.

Antimonii chloridum, F. 737.

Antimonii et potassæ tartras, F. 35, 256, 354, 538, 1095.

Antimonii vinum, F. 225, 253, 611.

Antiprurities. F. 233.

Apiol, F. 1054.

Argenti chloridum, F. 712.

Argenti ioduretum, F. 1087.

Argenti nitras, F. 149, 169, 200, 460, 469, 604, 650, 715, 739, 808, 815, 842, 857, 870, 883, 896, 913, 925, 963, 989, 995, 1022, 1031, 1035, 1083, 1140.

Argenti oxidum, F. 49.

Arnica, F. 429.

Aromaticum vinum, F. 935.

Arsenic, manner of administering, F. 223.

Arseniosi acidum, F. 115, 332, 340, 720, 725, 726.

Assafœtida, F. 2, 84, 87, 94, 95,, 1055 1137.

Atropia, F. 102, 1053.

Atropiæ sulphas, F. 64, 107, 109, 127, 128, 130, 197, 308, 393, 510, 874, 891, 1033, 1039, 1097, 1099, 1101, 1127, 1141.

Atropiæ valerianas, F. 1100.

Aurii et sodii chloridum, F. 93.

Belladonna, F. 24, 27, 39, 66, 108, 118, 120, 123, 124, 132, 135, 148, 188, 207, 300, 308, 387, 440, 441, 443, 447, 520, 558, 620, 831, 841, 900, 970, 1045, 1048, 1065, 1093, 1096.

Belladonnæ linimentum, F. 27.

Benzoin, F. 147, 157, 182, 565, 566, 567, 607, 695.

Bergamii oleum, F. 752.

Bismuthi subcarbonas, F. 19, 1074, 1075.

Bismuthi subnitras, F. 47, 48, 435, 450, 463, 466, 467, 556, 722, 813, 893, 1072, 1073.

Cimicifuga, F. 76.

Cinchona, F. 25, 56, 74, 207, 224, 276, 290, 292, 300, 315, 350, 351, 356, 377, 380, 404, 417, 519, 528, 605, 711, 854, 949, 953, 961, 967, 991, 1005, 1015.

Cinchoniæ sulphas, F. 74, 112.

Citrici acidum, F. 16, 462, 527.

Coccus cacti, F. 576, 655, 1094, 1095.

Colchicum, F. 28, 230, 511, 512, 515, 518, 519, 523, 524, 535, 536, 537, 609, 727.

Colocynth, F. 28, 34, 44, 57, 132, 423, 439.

Conium, F. 67, 108, 111, 139, 205, 208, 261, 275, 289, 293, 353, 657, 729, 1088, 1104, 1105, 1145, 1146.

Copaiba, F. 221, 278, 807, 808, 816, 817, 818, 819, 820, 824, 825, 827, 828, 845, 856, 860, 865, 868, 869, 901, 902, 903, 904, 905, 906, 907, 910, 911, 939.

Copaibæ aqua, F. 871, 872, 873, 874, 875, 876, 879, 876, 880, 881, 882, 884, 885, 886, 887, 888, 889, 892, 893, 894, 895.

Creasotum, F. 296, 316, 392, 461, 483, 615, 714, 719, 731, 733, 746, 771, 773, 775, 777, 796, 968.

Creta preparata, F. 144, 754.

Cubeba, F. 183, 260, 278, 807, 800, 817, 818, 821, 822, 823, 824, 825, 826, 828, 853, 865, 901, 902, 903, 904, 903, 909, 912, 939.

Cupri carbonas, F. 797.

Cupri sulphas, F. 167, 215, 404, 408, 454, 459, 560, 693, 866, 875, 877, 887, 1030, 1147.

Cuprum ammoniatum, F. 75, 724.

Daturia, F. 1106.

Digitalin, F. 1107, 1108.

Digitalis, F. 3, 61, 66, 220, 235, 279, 311, 341, 372. 377, 382, 383, 387, 389, 390, 391, 494, 502, 503, 621, 622.

Dulcamara, F. 294, 707.

Elaterium, F. 392, 496, 499, 500.

Emplastrum epipasticæ, F. 141.

Ergot, F. 137, 145, 220, 395, 396, 849, 1060, 1110.

Ergotin, F. 396, 1062, 1109.

Farina, F. 957.

Ferri ammonio-sulphas, F. 349.

--- ◆ - --- —

INDEX OF DISEASES.

26

INDEX OF AUTHORS.

406

www.ingramcontent.com/pod-product-compliance
Lightning Source LLC
Chambersburg PA
CBHW021348210326
41599CB00011B/792